THE WONDER OF WATER

Lived Experience, Policy, and Practice

Facing droughts, floods, and water security challenges, society is increasingly forced to develop new policies and practices to cope with the impacts of climate change. From taken-for-granted values and perceptions to embodied, existential modes of engaging with our world, human perspectives impact decision-making and behaviour.

The Wonder of Water explores how human experience – including our cultural paradigms, value systems, and personal biases – impacts decisions around water. In many ways, the volume expands on the growing field of water ethics to include questions around environmental aesthetics, psychology, and ontology. And yet this book is not simply for philosophers. On the contrary, a specific aim is to explore how more informed philosophical dialogue will lead to more insightful public policies and practices.

Case studies describe specific architectural and planning decisions, fisheries policies, urban ecological restorations, and more. The overarching phenomenological perspective, however, means that these discussions emerge within a sensibility that recognizes the foundational significance of human embodiment, culture, language, worldviews, and, ultimately, moral attunement to place.

INGRID LEMAN STEFANOVIC is dean of the Faculty of Environment and a professor in the School of Resource and Environmental Management at Simon Fraser University. She is also a professor emerita in the Department of Philosophy at the University of Toronto.

The Wonder of Water

Lived Experience, Policy, and Practice

EDITED BY INGRID LEMAN STEFANOVIC

UNIVERSITY OF TORONTO PRESS
Toronto Buffalo London

ISBN 978-1-4875-0593-6 (cloth)   ISBN 978-1-4875-3298-7 (EPUB)
ISBN 978-1-4875-2403-6 (paper)   ISBN 978-1-4875-3297-0 (PDF)

Library and Archives Canada Cataloguing in Publication

Title: The wonder of water : lived experience, policy, and practice /
   edited by Ingrid Leman Stefanovic.
Names: Leman Stefanovic, Ingrid, editor.
Description: Includes bibliographical references and index.
Identifiers: Canadiana 20190172835 | ISBN 9781487524036 (softcover) |
   ISBN 9781487505936 (hardcover)
Subjects: LCSH: Environmental ethics. | LCSH: Water consumption – Moral
   and ethical aspects. | LCSH: Water resources development – Moral and
   ethical aspects. | LCSH: Water-supply – Management – Moral and ethical
   aspects. | LCSH: Water conservation. | LCSH: Water – Social aspects. |
   LCSH: Water and civilization. | LCSH: Right to water. | LCSH: Water security.
Classification: LCC GB665 .W66 2019 | DDC 179/.1 – dc23

University of Toronto Press acknowledges the financial assistance to its
publishing program of the Canada Council for the Arts and the Ontario Arts
Council, an agency of the Government of Ontario.

Canada Council   Conseil des Arts
for the Arts   du Canada

ONTARIO ARTS COUNCIL
CONSEIL DES ARTS DE L'ONTARIO
an Ontario government agency
un organisme du gouvernement de l'Ontario

Funded by the   Financé par le
Government   gouvernement
of Canada   du Canada

Canada

Contents

Figures

Acknowledgments

I am pleased to acknowledge the financial support from the Social Sciences and Humanities Research Council of Canada, offered in the form of an Insight Development grant. The funds supported an early workshop and fundamental research, as well as preparation and presentation of the present book.

Special thanks are extended to our copy editor, Anne Louise Mahoney, whose invaluable assistance ensured a carefully reviewed set of chapters prior to submission to the press.

Similarly, I am sincerely grateful to editor Jodi Lewchuk at the University of Toronto Press, whose cheery, informed advice was a unique source of support.

I thank the authors whose work and thought constitute this book. I am particularly obliged to those who remained good-natured as the project was reframed during the research, causing some small delay in the preparation of the present volume. Their patience and support are much appreciated.

Reviewers who took the time to provide thoughtful feedback also deserve a word of thanks. It is never easy to find the time to commit to reading and evaluating another's work, and I am pleased to have had their support of this project.

I extend a word of thanks to members of the Dean's Office in the Faculty of Environment at Simon Fraser University, who provided me with ongoing professional assistance, allowing me to find time during weekends, vacations, and occasional brief research leaves to complete this book.

Finally, I am grateful to my family, and especially to my husband, Michael, who provide meaning and purpose to my life while inspiring me to continue to learn, grow, and thrive through my work. It is they who keep my days whole.

THE WONDER OF WATER

Lived Experience, Policy, and Practice

Introduction

There are these two young fish swimming along and they happen to meet an older fish swimming the other way, who nods at them and says: "Morning, boys. How's the water?" And the two young fish swim on for a bit and then eventually one of them looks at the other and goes, "What the hell is water?"

– David Foster Wallace

That our life-world is defined by water is not always evident. Many are unaware that over half the adult human body consists of water. Water defines us as embodied creatures – physiologically, but also culturally, socio-economically, politically, ontologically, aesthetically, and morally.

Today, climate change disrupts our relations with water, whether through drought or flood, or contamination of our rivers, lakes, oceans, and groundwater. Growing challenges arise around water security for communities worldwide.

The statistics are shattering: the World Health Organization reports that three out of ten people lack access to safe water at home; every year, over 350 million children die before reaching the age of five from diarrhoea linked to poor sanitation (2017).

These sorts of challenges demand changes in policies and practices at municipal through national and international levels. Often strategies for positive change are presented in technical briefs and reports that nevertheless miss the genuine meaning of water in its visceral quality, its vitality, and its primordiality.

This book aspires to move us beyond statistics and calculations, helping us to *see* water differently and behave more discerningly in respect of water. The phrase famously attributed to Socrates that wonder is the beginning of wisdom bids us ask: might a deeper, embodied vision of the wonder of water inspire more thoughtful policies? Could our built places be more wisely designed if we attended to water's lessons in a more meaningful way? In recalling the full depth of the lived experience of water, is it possible to rethink the meaning of water ethics, a new and growing field of study unto itself?

In exploring answers to these questions, each chapter contributes to a broader narrative that seeks to link philosophical reflection with best practices for decision making. Strictly speaking, from an academic perspective, the entries take a phenomenological approach to issues of water. The phenomenology movement arose as an alternative to reductionist, calculative modes of metaphysical reasoning. In place of analytic, theoretical constructs, phenomenologists seek to uncover taken-for-granted ways of being-in and knowing about the world in which we belong.

Despite the primordial, embodied significance of water, surprisingly little has been written by phenomenologists, not only about its existential meaning but also about the potential effect of such a way of understanding environmental decision making and public policies. Admittedly, the early twentieth-century phenomenologist Gaston Bachelard planted the seeds of such a conversation when he explored the oneiric significance of water through a phenomenology of the imagination. Describing water as "the transitory element," he acknowledged a special "type of intimacy" that defined our relationship to water (1983, 6). Others grappled with the implications of Goethean phenomenology, beautifully illumining the nature of "flow forms" and archetypal movements of water (Schwenk 1965).

But it is only very recently that the conversation around an embodied sense of the wonder of water might be said to have been taken up more consistently by philosophers and humanists. Sarah Allan movingly describes the water imagery found in early Chinese thought (1997). Christiana Peppard explores how theology and a conception of water as sacred can better inform decision making around the challenges of hydraulic fracturing, industrial agriculture, and climate change (2014). Astrida Neimanis and others aim "to rethink embodiment as watery," drawing from a feminist phenomenology to challenge three humanist approaches to corporeality: "discrete individualism, anthropocentrism and phallogocentrism" (Neimanis 2017, 2, 3; Chen et al. 2013).

Such books are rare – and often very recent – instances whereby writers begin to thoughtfully explore the embodied presence of water. More remains to be done, however, to deepen that conversation and to expand it to address related questions of water policy. While the current collection aims to delve more profoundly into the phenomenological discussion of the embodied experience of water, the goal is to move beyond the strictly philosophical to a more interdisciplinary discourse, one that will ultimately inform decision makers who are seeking a more thoughtful, discerning way of developing public policies around water.

Lived experience is complex; seeking to understand it means being open to its manifold manifestations and acknowledging that it exceeds the limits of simplistic, reified categories. Narrowly mathematical reasoning needs to be supplemented by philosophical reflection that incorporates all manner of emotional, aesthetic, cultural, socio-economic, political, and ontological ways of being in the world. In this volume, phenomenological discourse reveals a broad spectrum of such ways of engaging with water specifically. That discussion often emerges within the context of the growing field of environmental and architectural phenomenology, a field that promises to enlarge our understanding of both natural and built places.[1]

However, one need not be a phenomenologist to benefit from this book. The collection is meant to move beyond an academic audience, drawing upon philosophy to inspire not only phenomenologists interested in further reflecting on water issues, but also decision makers who engage with water policies and practices on a daily basis.

To that end, this book is composed of four parts:

- Part One recalls the phenomenological roots of "The Lived Experience of Water." Often taken for granted, our relation to water is described not simply in a cerebral, abstract sense, but rather as it occurs through embodied encounters.
- Part Two explores the relation between "Water and Place," recognizing that water both defines who we are, but also where we are implaced.
- Part Three, "Rethinking Water Policy, Practice, and Ethics," builds on the earlier chapters to seek guidance about how to develop more thoughtful water policies and practices.
- Part Four presents "Closing Reflections."

In Part One, the authors explore what it means to situate water within the context of our everyday embodied "lived experience." Poet Kirby Manià opens the book with her brief but moving reflections on the wonder of reading water landscapes by way of such lived experience. The fact that the contents of this volume are framed, beginning and end, by poems serves to remind us that the poetry and wonder of water invite more than a technical set of policy solutions and technological fixes.

Writing as both a scientist and a phenomenologist, **Stephan Harding** builds on this vision by engaging in an "imaginative visualization" of cosmic water. The depth and breadth of our planet's ability to retain life-giving wetness invites us to reflect on water's vastness, both spatially and temporally – and its wholeness in *Gaia*. This chapter is less an experience of reading *about* water than it is imaginatively reliving, through Harding's deeply moving narrative, its foundational, life-giving, cosmic origins.

Phenomenologist and health educator **Stephen J. Smith** explores our embodied, kinetic relations to water and its flow motions. Identifying what he calls "waterscape gestures" brings new meaning to the essential attunement with the more-than-human world that situates lived experience. From the desperate need of water known through moments of thirst to the fury of storm events, Smith leads us to a rethinking of water ethics as deeply attuned to life itself. Written less as an analytical argument than a poetic discourse, Smith's own chapter itself reflects a watery flow.

David Abram similarly draws from his phenomenological background, situating the discussion of water within the context of other-than-human living beings. His descriptions of life's migratory draws and the allure of seasonal memories reveal the visceral imperatives of our living, breathing world and its waterways.

Philosopher **Martin Lee Mueller** then draws Part One to a close, contributing some initial thoughts about the themes of the sections to come – place, policy, and practice. This is a rich chapter, questioning the tendency to situate the salmon industry within dominant and domineering calculative paradigms. From the Pacific Northwest to Norwegian examples, Mueller describes our shifting relations and the choices we make with respect to salmon and to larger waterscapes.

Part Two moves to a discussion of "Water and Place." Our lived experiences do not occur within a vacuum. To be is to dwell as implaced. Phenomenologist **Janet Donohoe** shows how water is often taken for granted as a fundamental element of dwelling itself. Helping to inspire the reader to become "more attentive to water in its waterness," she aims to show that when our placemaking is brought into line with water, the result is more thoughtful planning and policy making.

Irene Klaver enlarges upon that conversation about placemaking by focusing on examples drawn from Amsterdam and the Netherlands as a whole. The meandering dimensions of "riverspheres" help to define conflicts and opportunities that are deeply social, political, cultural, and ecological in their import. They also offer the chance to rethink the meaning of agile planning and democratic policy making in Europe and beyond.

How water defines city building continues to be explored by the editor of this volume, **Ingrid Leman Stefanovic**. Water infuses cities – defining, sustaining, and transforming our sense of place. Two narratives of place – illustrating, in one case, water's fluidity, and in the other, the power of stillness – reveal how architectural design opens up a space to rethink anthropocentric worldviews, acknowledging instead that we are embedded "within a watery world whose ontological givenness we did not create."

Sarah J. King closes Part Two, using the case of Flint, Michigan, as an illustration of how race, imperial politics, and environmental ruination define "What We're Talking About When We're Talking About Water." The poisoning of one American city's water system is seen to have arisen within the context of underlying problems such as racism, imperialism, marginalized communities, and the power structures that keep them in place. Drawing comparisons with the fishing disputes at Esgenoôpetitj/Burnt Church, New Brunswick, King illustrates how water conflicts often reflect relations of power that do deep violence to communities' sense of place.

King's case studies help us to transition to Part Three, "Rethinking Water Policy, Practice, and Ethics," which focuses on praxis, policies, and decision making about water. **Bryan E. Bannon** begins by reflecting on how occurrences of "daylighting" underground streams can sensitize us to a restorative relation to water and the natural world. Advocating a phenomenological approach to meaning as inhering in such relations rather than in the properties of individual entities, he shows how more democratic and sustainable practices emerge when we assume an attitude of "friendship" towards the world.

Trish Glazebrook and **Jeff Gessas** also draw on a case study, that of Standing Rock, to explore "Water Protectors in a Time of Failed Policy." The chapter recounts a story of the Lakota Sioux and their efforts to protect the waters of Lake Oahe from the negative effects of a pipeline. A gathering that was arguably the largest assembly of Indigenous Americans in over 150 years is seen by the authors to have been a historic, game-changing rupture of the predominant logic of water, by a phenomenology of water. Drawing on an ecofeminist perspective, the authors show how an Indigenous worldview offers the promise of "living-with water in a holistic, community-driven way of life that values people across generations, other species, and ecosystems."

Henry Dicks weaves together philosophy and praxis through his discussion of "Phenomenology, Water Policy, and the Conception of the Polis." Dicks argues that how we determine water *policy* is grounded principally in how we conceive the *polis*. He suggests that depending on whether we frame our understanding of the polis by analogy with the human body, an organism, or a forest ecosystem, "water will manifest itself in fundamentally different ways." Different water policies, therefore, will emerge.

How phenomenology can inform a new ethic is the topic of the chapter by **Robert Mugerauer**. Inviting us to move "Towards a Complexity Ethics," we again draw lessons from a case study, this time of the Lower Duwamish Waterway. Acknowledging that ethical concerns blend into politics, public policies, and practices, Mugerauer argues that water ethics, to be meaningful, must arise from within an ethic of care and a sensitivity to the complexity of lived networks of relationships, if so-cial-ecological well-being is to be achieved.

The book concludes with a selection by poet **Dilys Leman**. Closing with her poetic series around Toronto's own Don River reminds us of how our daily lives are always situated against the backdrop of water and waterways. Concluding remarks are offered in both a spirit of closure – reflecting on lessons offered by this uniquely thoughtful set of authors – as well as an open invitation for each of us to continue to see water, feel it, experience it, remember it, and never take it for granted as we continue to generate policies and practices to protect it.

NOTE

1 See, for instance, the Environmental and Architectural Phenomenology Network Newsletter, edited by David Seamon, at https://newprairiepress.org/eap/.

REFERENCES

Allan, Sarah. 1997. *The Way of Water and Sprouts of Virtue.* Albany, NY: State University of New York Press.

Bachelard, Gaston. 1983. *Water and Dreams: An Essay on the Imagination of Matter.* Dallas: Pegasus Foundation.

Chen, Cecilia, Janine MacLeod, and Astrida Neimanis. 2013. *Thinking with Water,* Montreal & Kingston: McGill-Queens University Press.

Neimanis, Astrida. 2017. *Bodies of Water: Posthuman Feminist Phenomenology.* London: Bloomsbury Publishing.

Peppard, Christiana Z. 2014. *Just Water: Theology, Ethics and the Global Water Crisis.* New York: Orbis Books.

Schwenk, Theodor. 1965. *Sensitive Chaos: The Creation of Flowing Forms in Water and Air.* East Sussex, GB: Rudolf Steiner Press.

World Health Organization. 2017, 12 July. "2.1 Billion People Lack Safe Drinking Water at Home, More Than Twice as Many Lack Safe Sanitation." News release. Accessed 10 October 2017. http://www.who.int/mediacentre /news/releases/2017/water-sanitation-hygiene/en/.

PART ONE

The Lived Experience of Water

In many ways, one might expect a book like this one to invite the reader to confront issues of water ethics and policy in an explicitly deliberative, calculated manner. From debating the morality of bottled water to developing policies to deliver safe water to remote, Indigenous communities, contemporary conversations are often structured in a way that intellectually delimits and categorizes options as legibly and rationally as possible.

Certainly, discussions of water policies and decision making should not flout the rules of reason. That said, life is messy. Policies and practices are often not established along narrow parameters of theory or logic; nor should they necessarily be. Rather, they can only benefit from the wisdom that emerges from a more holistic understanding of the full range of lived experience that defines human engagement in our daily activities.

Part One of this book reveals the significance of lived experience in decision making on issues of water policies and *praxis*. The notion of lived experience has its roots in Husserlian phenomenology and his conception of the life-world (*Lebenswelt*). Phenomenology as a movement was a response to a long metaphysical tradition of dualistic thinking that distinguished between and theoretically separated subjects and objects: mind and body; science and art; metaphysical contemplation and the physical world. Husserl sought a different way forward, to uncover the initial belonging and taken for granted context that constituted the condition of the possibility of the dualistic bifurcation of elements of experience in the first place.

In confronting the reifying tradition of metaphysics, Husserl was able to demonstrate how "we cannot conceive subjectivity as an antithesis to objectivity, because this concept of subjectivity would itself be conceived in objective terms. Instead, transcendental phenomenology seeks to be 'correlation research.' But this means that the relation is the primary thing and the 'poles' in which it forms itself are contained within it" (Gadamer 1975, 220). In fact, the term "lived experience" is meant to indicate that we function not simply as cerebral, intellectual, reasoning creatures but that our ways of interpreting and understanding the world also include memories, values, culture, language, emotions, and taken-for-granted, embodied ways of engaging with the world as well. "The concept of the 'life-world' is the antithesis of all objectivism ... by the life-world is meant something else, namely, the whole in which we live as historical creatures" (Gadamer 1975, 218).

Acknowledging the significance of lived experience also means recognizing that our ways of knowing the world are not simply constituted by explicitly known, calculative objects or *things*. Rather, the pre-thematic horizon of understanding within which those things appear is included within the realm of what we mean by "lived experience." Whether a hammer that is put to use or a stick that guides the blind man, or the eyeglasses through which we see the world, each such tool becomes a part of my lived experience without necessarily appearing as an object in and of itself: as elements of our daily life, such items "'withdraw' and are barely noticed, if at all" (Idhe 1990, 73). Lived experience includes both the thematic as well as the pre-thematic relations and context within which objects find their place.

Each author in this section helps to advance our understanding of the significance of lived experience in relation to our conversations around water. **Kirby Manià's** poem, *Rain Queen*, reveals how the language of our natural, watery world is not only cerebral but is deeply embodied. The lived world "always implicates the body-in-action" and indeed, how we read the skies and the landscape with the approaching rain reflects that essential embodiment (Ihde 1990, 39).

Stephan Harding's chapter immerses us in the originary, live-giving cosmic dimensions of water that, despite being essential to the world as we know it, so rarely enter into conversations about policy making and praxis, even though they serve as the essential, tacit context for these very conversations. Moving from Harding's imaginative visions of the planetary dimensions of wholeness in *Gaia*, **Stephen Smith** then invites us to explore our own ancestral, kinetic affinities with water and its "primordial amniotic power." Smith's reflections show how human consciousness is integrally linked to waterscapes that invite us to contemplate our essential belonging to the more-than-human ways of being in the world.

David Abram similarly takes us beyond the narrow parameters of anthropocentric discourse, situating us within the larger, lived experience of migratory patterns of the natural world, and the animals and waterways within it. Abram's chapter serves to remind us that when we slip into our instrumental discussions of water as a resource or commodity, it is important to look up and around us to see how those conversations are meaningfully situated within the patterns of natural movement and change that define our existence.

Martin Lee Mueller closes this section with his reflections on the salmon industry and how policies and practices can be all the more meaningful if they move beyond our western, engineering paradigms to a more thoughtful engagement with the non-human animal world. By rooting his discussion in specific case studies from the Pacific Northwest to Norway, he also opens the door to the theme of Part Two: the significance of place.

REFERENCES

Gadamer, Hans-Georg. 1975. *Truth and Method*. New York: The Seabury Press.
Ihde, Don. 1990. *Technology and the Lifeworld: From Garden to Earth*. Bloomington and Indianapolis: Indiana University Press.

Rain Queen

As a child,
I always wondered
how my parents could tell
there was rain
on the horizon
their adulthood rendering
authorship to landscape,
thereby able
to read the language of the sky.
To me,
it was as unfathomable
as their alphabet –
the cloud's page silent
in its mastery
over a world that seemed
so vast and wild
in its unknowability.
But once grown,
I have since learnt
how to speak
this tongue –
sight reading
the brooding firmament's
True Name
It is a world
now sadly
curtailed in wonder
and majesty –
grown-ups' empiricism
bereaving
the literate.

Kirby Manià

1 Water Gaia: Towards a Scientific Phenomenology of Water

STEPHAN HARDING

Water is very old. Not, of course, the water you are breathing out now as you read this page. No – that's new water, a fresh reminting of two hydrogen beings with one of oxygen made by that oxygen-fuelled slow burn of food in your cells that we call metabolism. Old water has been around in the cosmos in vast quantities for a long time, long before our solar system existed. The oldest water was created at least 12 billion years ago, during the early life of our cosmos just after the explosion of the first supernovae – those huge stars in whose innards all the chemical elements heavier than hydrogen are created. Thanks to these massive exploding stars, we live in a water-spangled cosmos where about 10 per cent of all the matter in the vast reaches of space is water in the guise of water ice.

Since we can't see this old cosmic water with our naked eyes, perhaps the first step in our scientific phenomenology of water requires us to engage in the practice of what I call imaginative visualization.

Imagine, then, the cosmos shortly after the supposed big bang.

Visualize, if you will, in whatever way you can, vast clouds of hydrogen, the primordial element created in the big bang itself, smeared throughout space as a diaphanous tissue of shifting gaseous vapour, clumping here and there into denser pockets due to gravitational attraction. Now visualize some of these clumps aggregating down to such huge densities that the pressure inside them causes the hydrogen atoms to fuse into helium atoms, emitting vast amounts of light in the process. Thus are stars born.

Visualize now how in some of these bigger stars, the inward pull of gravity among their helium and hydrogen atoms causes further fusion, giving rise to many of the heavier elements, including oxygen. Visualize these huge stars exploding as supernovae when their inward pressure becomes so immense that a kind of massive nuclear blast is created, spewing out the heavy elements into the surrounding space.

Out here, oxygen and hydrogen atoms can meet each other in more peaceful surroundings. See them now, finding fulfillment by sharing electrons with each other, creating a new emergent molecular being with the unique, unexpected qualities of that slippery, sometimes liquid H_2O we call water. If, as some of our greatest philosophers have intuited, the cosmos is a great psyche, then water must be one of its most ancient *ideas*. Seen thus, each water molecule is a primordial concept made real due to the elemental attractions between hydrogen and oxygen.

We could keep on visualizing old water in the distant past out in space, but since we are aiming for a more current phenomenology of water, let's now shift our focus to water right here on our planet, on Earth, Gaia (Lovelock 1995). Where did our water come from? Here we encounter two possibilities, not necessarily mutually exclusive. The one with the most mythological appeal involves the gas giants Saturn and Jupiter. In essence, the idea is that early on in the life of our solar system, the orbits of these two huge planets shifted, sending some huge chunks of water ice from the asteroid belt hurtling towards Earth and the other inner rocky planets. It would have taken perhaps only a few of these icy comets to give us all the water now on Earth.

In my more poetic, less scientific moods, I like to imagine Father Jupiter peering in at the inner rocky planets of our solar system, and, feeling sorry for their desiccated state, using his gravitational power to send those comets in Earth's direction. Some must have missed, plunging into the sun with colossal puffs of steam. Others would have hit Mercury, Venus, Earth, and Mars, giving them all their primordial water. The other, more prosaic, hypothesis is that the water was there in the very grains of cosmic dust that slowly accreted to form Earth and her rocky neighbours. There's more evidence for the latter hypothesis, which, sadly, shows that evidence doesn't always confer with poetic intuition.

Now that we've used imaginative visualization in our scientific phenomenology to discover how water in the cosmos was created and how some of that water might have ended up on Earth, it's time to explore how water behaves at the temperatures and pressures it experiences here at our planet's surface. It's time for an experiment, or as the phenomenologically correct French would say, an *expérience*. Put some tap water into an ice cube mould and place this in your freezer until you've made some ice cubes. Now you have ice – solidified water. Next, drop some of these ice cubes into a glass of water, and carefully observe the startling result. The ice floats! The exclamation mark is essential, because this really is a most astonishing phenomenon that has massive implications for life on Earth. Most substances become denser as they cool, which means that their solid phases (ices) sink into their corresponding liquids. Ice is less dense than water, so it floats.

Imagine that we could work some devious chemical magic so that the entirety of the world's ice suddenly became denser than water. The ice would sink, and many areas at the bottom of the oceans would soon be covered in thick layers of ice, making life impossible for bottom-dwelling creatures such as kelp forests, crabs, and many sediment bacteria. This same sediment would lose its contact with seawater, so ocean currents would not be able to carry nutrients such as phosphorus in the sediment to the ocean surface. This would starve the photosynthetic plankton of such nutrients, the foundation of the entire marine food web. So, no fish, no whales, and so on, and no carbon fixation from the atmosphere by the phytoplankton, so the global temperature would increase substantially. There would be further catastrophes. Imagine the polar regions no longer sporting their reflective sea ice – this, too, would make the planet warmer due to the loss of the ice's sun-reflecting surfaces. Without sea ice to keep them in place, ice sheets currently grounded on Greenland and on Antarctica would slide into the sea, raising both sea levels and global temperatures.

Ice floats on water because of a particularly special relationship between the hydrogen and oxygen atoms in water molecules – this is the key to many of its astonishing properties. Let's embark on another scientific phenomenological visualization. This time we shrink ourselves to 3×10^{-10} m, small enough to see individual water molecules as we plunge into a glass of water. We see trillions of them in movement, each one atom of oxygen and two of hydrogen arranged in a wide V-shape at 104.5°C, with the oxygen at the sharp end. Look carefully and you'll see that most of the eight electrons shared by the oxygen and hydrogen atoms spend most of their time zooming around the electron-hungry oxygen atom, which therefore has a stronger negative charge than would be expected if the electrons had been evenly shared. Correspondingly, the two hydrogen atoms emanate a slight positive charge. This seemingly insignificant charge differential within each water molecule is the secret to water's unusual behavioural qualities. The endless mystery of attraction between opposites, between positive and negative electrical charges, that elusive Eros that underlies so much of nature throughout the entire cosmos, now makes itself felt here at the tiny, submicroscopic scale of a single water molecule. See now how the more positively charged hydrogen atoms on the molecule we are observing are attracted to the negatively charged oxygen atoms on neighbouring molecules of H2O. This is the famous hydrogen bond – an attraction nowhere near as strong as the covalent bonds that hold the oxygen and hydrogen atoms in every water

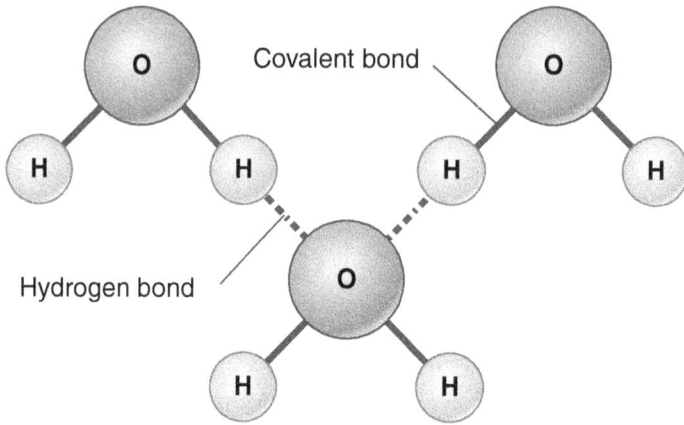

Figure 1.1. Hydrogen bond

molecule in that tight, more enduring embrace of atomic marital satisfaction, yet strong enough to give water its particular properties, which, as we will see, make life possible (see Figure 1.1).

Because they are so weak, hydrogen bonds form and break easily in liquid water in which water molecules, agitated by heat from their surroundings, gyrate and tumble over each other in an endlessly chaotic choreography that keeps them dancing wildly – as long as their temperature lies somewhere between water's very own freezing and boiling points. Now let's return to what happens to water's hydrogen bonds when water freezes to ice. With less heat to scramble them, the hydrogen bonds lock into permanence, creating interlinking hexagonal sheets of water molecules more widely spaced than those in liquid water. At last we see why ice floats on water – it is simply less dense than water because its hydrogen bonds are spaced more widely apart (see Figure 1.2).

However, at room temperature the hydrogen bonds, although weak, keep the water molecules bound together in the liquid state.

This liquid state – I see it clearly now as I sit on a little island at the edge of the River Dart, lost in contemplation. I connect with trillions upon trillions of tumbling water molecules, making and breaking their hydrogen bonds, becoming one emergent flowing whole: the river itself with its speech as it cascades over rocks – a synergistic duet of rock and water. I try to catch what it says – to parse that deeper voice I sense within it, but I never quite catch the message. The whiteness of the water as it roars past the rocks is another emergent property of those watery triplet beings making and breaking their myriad hydrogen bonds. So is the complex riffled surface I see as the river flows on downstream, carrying

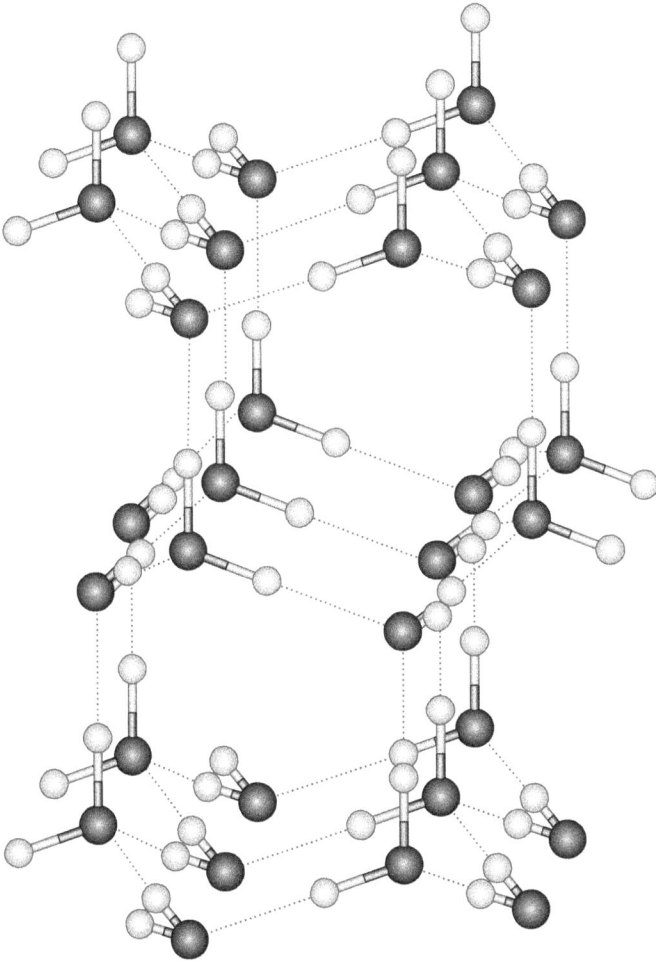

Figure 1.2. Snowflake structure

things I cannot sense but know are there, held in its watery embrace: fine sediment eroded from the naked fields, brown particles of earth. All sorts of ions – charged chemical beings – calcium, iron, magnesium, and many more, are dissolved in the river's water, each escorted by ragtag gaggles of water molecules attracted by the ions' electrical charges, giving them both freedom and protection.

   In part, it is water's ability to hold ions in this way due to the polarity between the slightly positive hydrogen atoms and the more strongly negative oxygen atom that makes it essential for life. Water's polarity means that it is an excellent solvent, ably and easily dissolving a huge range of substances needed by living beings into a water-bound liquid nutrient that is easily transported across cell membranes into the depths of the cellular interiors of living beings, where they are desperately needed. Water also helps to moderate the overall balance between the myriad charged molecules within the cell and dissolves wastes that the cell can easily excrete across its membrane. Also, water is liquid at temperatures that allow it to be a flowing matrix supporting the complex biochemical activities of living beings. If the temperatures allowing this liquid state were any warmer, these biochemicals would fall apart – any colder and they'd freeze.

To get a felt sense of how important water is for life, consider this: living beings are mostly water. We humans are 60 per cent water. Some other living beings, such as water plants, contain as much as 95 per cent water. As Vladimir Vernadsky, the great Russian pre-Gaian scientist of the early twentieth century, once said, "Life is animated water." For scientists Mark and Dianna McMenamin (1994), life on land is one huge interconnected cellular meshwork of flowing water which they call "Hypersea" – literally, a living, pulsing interactive sea within the tissues and cells of all land organisms whose ancient ancestors engulfed the ocean into their bodies when they first colonized the land. The inter- and intracellular waters of Hypersea move horizontally through soils within fungal tubes and into the plant roots with whom they connect symbiotically, sharing nutrients and information, promoting diversity in Gaia's terrestrial ecosystems. Every land organism carries its own personal piece of this flowing Hypersea within its cells and among its tissues, gifting its precious fluids to others when eaten, parasitized, or dying.

For Italo Calvino (1976) and Andreas Weber (2016), this ocean that we carry within us, this Hypersea, gives us the chance to feel more deeply into life – to experience the ocean, in poetic imagination, as a great inside where life originated and which we now carry as another inside here in our very own bodies. This notion of "inside" suggests that both ocean and Hypersea possess some kind of sentience, some kind of mind. The ocean isn't just water, it feels and knows, as do the sea-like body fluids that saturate our every cell. When I swim in the ocean, says Weber, my inside meets and feels that far greater other, which is also an inside.

I sneak through sedges, docks, and pink campions in my wild garden, lie on my stomach and place my eyes close to the water surface of the little pond I dug in here twenty years ago. Tea-dark water, quite clear, shaded now from the bright spring sunshine by overhanging brambles and the old ivy-covered wall. The pond world slowly opens itself up to my gaze as a newt swims languidly to darker depths and tiny *Daphnia* beat the water with their antennae. Rotten leaves on the bottom, threaded here and there by rooted, undulating worms. Beneath my gaze, slowly moving flatworms hang upside down directly on the underside of the water surface. If I did the same, I would lie spread out on the ceiling of my sitting room, my back facing the floor. How can the flatworms do such a thing? Zoom down with me once again to the scale of a water molecule, right into the very surface of the pond water. See how the water molecules here have fewer partners to bind with, because there are none in the air above? This means that these particular water molecules cling more tightly to each other via their hydrogen bonds, creating a zone of tense water that can support the weight of an upside-down flatworm below or of a splay-legged water skater above. This same surface tension pulls water into limpid drops when it free falls from leaf tips or is splashed out from waterfalls. It seems increasingly likely that surface tension might be created, in part at least, by a fourth phase of water known as exclusion zone, or EZ, water, in which coherent liquid crystalline strands of water molecules form around water-loving surfaces and at the surfaces of water.

These hydrogen bonds – have I done well enough in giving you a felt sense of how important they are in making water what it is? If not, let me illustrate further. Let's engage in another chemi-magical experiment. With a wave of the hand we replace each and every oxygen atom in our glass of water with sulphur atoms, oxygen's immediate down-column neighbour in the periodic table. We've alchemically converted our water into H_2S – hydrogen sulphide, which makes no hydrogen bonds because electrons don't find sulphur atoms particularly attractive. Without these bonds to bind them, the hydrogen sulphide molecules rapidly pick up ambient heat and, gyrating themselves into a frenzy as a gas, immediately vapourize from the glass, saturating the surrounding air with their signature highly toxic rotten egg aroma. Some numbers might help here for developing a phenomenological feel for the subtle strength of water's hydrogen bonds. Hydrogen sulphide is liquid only below an exceedingly cold −60°C. Any warmer and it boils into gas. Water boils at a scalding 100°C (at sea level, of course). This huge difference of 160°C is due entirely to the cohesive energy of water's hydrogen bonds. Can you feel their power?

Boiling is fast-track evaporation – a release of gaseous molecules from the surface of a liquid: in our case, of water. On my way to the river through the woods, I looked up beyond the trees and saw *white sky water* – clouds – mostly water evaporated by the sun from the nearby ocean. Clouds are one of water's life forms. How they hang in white puffs or extended sheets in the sky is a mystery to my animal body, but my rational mind is poised, ready to offer its story – several, in fact – to explain this mystery. Here is one. Plants suck up water from the soil with their roots, sending it up their stems with the help of surface tension through hollow tubes as far as their leaves, on the undersides of which are tiny pores through which the water escapes into the air along with specialized chemicals crafted by the plants to condense the water into clouds up in the atmosphere. Water thus condensed can fall as rain, returning the water to earth, soil, and vegetation.

Cloud seeding takes place on a massive scale over large areas of forest such as the Amazon. Here, huge numbers of cloud-seeding chemicals made by living beings, mostly trees, chase after water molecules wafting into the air from the forest itself. The airborne water molecules condense around these chemicals in vast numbers, forming clouds. The conversion of water molecules from gas (water vapour) into cloud lowers the air pressure above the forest simply because there are fewer gas molecules left in the air when a cloud forms. Nature abhors a vacuum, and so this low-pressure region is rapidly filled by warm, high-pressure air full of water vapour literally sucked in over vast distances from the faraway ocean. The result is a huge aerial river of moisture-filled rain producing air surging all over the forest from the distant sea. Thus does the forest create and recycle its own rainfall. This is just one strand of the great planetary water cycle in which water evaporates from the ocean, travels far and wide in currents of air, and falls as rain over the land, sculpting canyons and river valleys as it goes, eroding hills and mountains, finally finding its way back to the ocean via rivers carrying sediment and organic material to the sea floor.

There is more. It seems likely that some bacteria are sucked up into clouds via strong updrafts of air created when clouds form over the planetary surface. These bacteria use clouds as a means of dispersal, much as fungal spores use the wind. There is evidence that some kinds of bacteria hitching rides in clouds trigger ice formation by secreting ice-nucleating compounds onto their external surfaces. The ice, loaded with bacteria, falls through the cloud as rain or snow, depositing the bacteria in pastures new. I see clouds very differently since I learned about these astonishing findings. No longer are clouds the result of mere physics. They are Gaian phenomena created by intricate and surprising relationships between living beings and those humble yet climatically powerful water molecules (Harding 2009).

Rain falls as I walk back from the river through the woods. At first, a few tentative drops, then, very quickly, a deep hissing rain, drenching leaf and soil in minutes. Life-giving water falling from the sky, pelting onto leaves, forming thin, flowing membranes that gather into drops at numberless leaf tips, glowing like diamonds, each a mass of zooming water molecules held together by its own transparent bag of surface tension. The leaves convert the violent pelting into a slow cascade of water drops that fall to earth like kisses, vanishing like ghosts into the interstitial spaces of the soil. Some of these water molecules, as we have seen, are captured by the tree's roots to be sent to the leaves where some depart intact to the air. Others are captured by the intricate quantum entanglements of photosynthesis in the leaves, which uses sunlight to split water molecules into solar-energized electrons and atoms of oxygen and hydrogen. These hydrogen atoms, together with carbon dioxide molecules inhaled from the atmosphere, are assembled by the plant into wood, sugars, DNA, and all the plant's many biomolecules using solar power carried by the energized electrons. The oxygen is exhaled via the leaf pores into the surrounding air – the very oxygen that fuels the energy production systems of many an organism great and small, from bacteria to great whales, keeping their awareness active and alive, wherever they are. For the last 2,500 million years or so, ever since one single tiny bacterium supposedly invented it somewhere at the surface of the ancient sunlit ocean, this water-splitting, oxygen-producing photosynthesis has captured the solar energy that runs much of our planet's biosphere.

Life not only contributes to cycling water around the planet – it also helps to keep the planet moist by preventing water from escaping into space (Harding and Margulis 2010). There are many ways in which water can be lost from the planet, most of which involve splitting water into hydrogen and oxygen. Hydrogen is extremely light – so light, in fact, that our planet's gravitational field can't keep hold of it, even in its dihydrogen molecular form of H_2. If you want to get a sense of just how light hydrogen is, think of how it was used to lift large masses of metal, wood, and fabric in those old-fashioned airships like the *Hindenburg*.

I think of hydrogen as the teenager of the chemical world. As the first-born of the chemical elements, it wants nothing more than to escape from Earth into space, where it can join swirling gangs of other hydrogen molecules, rather like teenage humans gathering in shopping malls and town centres throughout the westernized world. So in which ways can water be lost from the planet, and how does life help prevent this?

Let's take a journey down to basalt rocks on the sea floor at the bottom of the ocean. These rocks are full of iron, which feels a powerful chemical-erotic attraction to oxygen, the passionate Italian of the chemical world. Oxygen and carbon dioxide dissolved in seawater react with iron held in ferrous oxide in the rocks, producing rust-coloured ferric carbonate and odourless hydrogen gas. So powerful is the attraction between iron and oxygen that it can truly be said that rust is nothing less than lust as it manifests in the soap opera of the chemical world. Since less hydrogen means less water, for every water molecule that takes part in this particular chemical liaison, one hydrogen molecule is set free to make its escape bid to space, thereby potentially drying out the planet one hydrogen molecule at a time. Living beings also liberate hydrogen from water, including undertaker bacteria that ferment dead bodies in the dark depths of bogs and sediments out of the reach of oxygen. However, thanks to other kinds of organisms, very little hydrogen manages to escape to space. A different sort of bacteria living on the sea floor – known to science as anaerobic chemoautotrophic bacteria – capture the fleeing hydrogen and combine it with dissolved carbon dioxide, making the organic components of their bodies and reconstituting water by combining two parts hydrogen with one part oxygen within their cells. Other bacteria capture hydrogen via different pathways, but for the last 500 million years or so, oxygen released from abundant photosynthesis to the atmosphere has chemically combined with escaping hydrogen, thereby reconstituting water. Furthermore, the main planetary temperature-regulating feedback on Earth, in which water and life are centrally involved in weathering silicate rocks such as granite, has kept the upper atmosphere cold enough to freeze out water molecules before they can be split higher up in the atmosphere by solar ultraviolet light. Thanks to these and other of life's activities, our planet now has some 100,000 times more water than there would now be on a contemporary Earth devoid of life.

To appreciate just how significant is life's ability to keep earth moist by capturing hydrogen, consider Venus and Mars. Both are now almost totally desiccated despite the fact that each had abundant water soon after the solar system was formed some 4.5 billion years ago. Venus, being closer to the sun than Earth, lost its water billions of years ago when intense solar energy evaporated all its ocean water, solar ultraviolet radiation splitting it into oxygen and hydrogen. The hydrogen, too light to be held by Venusian gravity, drifted off into space. When Mars was young, it hosted an ocean one and a half kilometres deep that may have covered 20 per cent of its surface. But by around 3 billion years ago, 87 per cent of that water had disappeared due to powerful charged particles from the sun splitting water, leading hydrogen to be lost to space. We can't directly experience life's ability to keep hold of water, but we can at least visualize hydrogen molecules being captured in various ways by bacteria and by life-produced oxygen here on Earth before they manage to escape into space. We give thanks to life for keeping our planet hospitably lush and moist.

There are further reasons to be grateful for the retention of wetness by life on our planet, for without water there would be no granite continents, no mountain chains, no deep oceanic subduction zones, no volcanoes, and possibly no magnetic field. The process that links these aspects of our planet is plate tectonics, which could not operate without water. When hot basaltic rocks from deep within Earth's mantle come to the surface at the mid-oceanic ridges, they spread sideways and absorb water as they go. Water softens these huge slab rocks, known as tectonic plates, making them more able to bend, fold, and fracture when they collide with each other at plate boundaries. Here, the more water-rich plate is pushed below the drier plate, creating a subduction zone. There are different kinds of subduction zones, but all depend on the rock-softening effects of water for their existence. Deep within some kinds of subduction zones, the water-rich basalt rocks on the sea floor experience temperatures and pressures intense enough to cook them with their water into a less dense, rocky porridge of molten granite, which rises to the surface, cooling as it goes, eventually solidifying into granite rock – the foundation of all the continents. This relationship, in which water is central, gives us a deeply Gaian relationship to contemplate: no life, no water; no water, no plate tectonics; no plate tectonics, no continents. If this is correct (and many scientists who know about these things can produce evidence that it is), then without life's ability to keep the planet moistened with water, there would be no Americas, no Africa, no Eurasia, no Australia – in essence, no large continents at all.

The impact of plate tectonics might go even deeper, to the very core of the Earth, for it could play a role in generating Earth's magnetic field, which deflects charged particles from the sun (known as the solar wind) that would otherwise strip our atmosphere away. The field itself arises in the outer core of the Earth, some 2,890 kilometres beneath your feet, a region of liquid iron about 2,300 kilometres thick. This liquid iron is white hot – around 4,400°C in its outer regions and 6,100°C towards the inner core. The magnetic field is generated by a dynamo effect when this liquid iron circulates in a convection cycle within the outer core, and it is possible that this is entrained by another convection cycle in the overlying semi-viscous mantle rocks. The idea is that these upwelling mantle rocks take heat away from the top of the outer core where the two rock types meet, thereby facilitating the circulating dynamo effect within the entire outer core. Since convection in the mantle requires plate tectonics, which needs water, it is at least conceivable that without water (and life), the Earth would have no magnetic field and therefore no atmosphere. This, for me, is a profoundly Gaian possibility: that the thin smear of life on our planet's surface, with its multifarious ways of capturing water, influences vitally important tumultuous goings-on in white-hot molten rocks thousands of kilometres in Earth's depths, far below our everyday world.

Back on the planetary surface, water also influences climate. The journeys of individual water molecules in and out of the biosphere, ocean, and atmosphere in the water cycle have hugely important effects on Earth's temperature, thanks to the ways water absorbs, reflects, and redistributes energy from the sun. Water can either warm or cool the Earth, depending on which of its phases (ice, liquid, or vapour) it finds itself in. Water vapour is the most important greenhouse gas, warming the Earth, whereas low-lying clouds of condensed water vapour cool by reflecting solar energy back to space. As we've seen, ice and snow also cool. Other, higher clouds – those wispy mare's tails, cirrus clouds – produce an overall warming effect.

The ocean, in its vastness, also has huge impacts on climate. It absorbs about 1,000 times more heat energy than the atmosphere, taking up 80 to 90 per cent of the heat trapped in the atmosphere by greenhouse gases emitted by our activities. Surface ocean waters warmed in the tropics travel to the high latitudes, making these regions warmer than they would otherwise be. I am a grateful recipient of this watery bounty, for Britain is warmed in winter by the Gulf Stream, which carries balmy tropical waters from the Gulf of Mexico to our shores. Without the Gulf Stream, winter temperatures here would be similar to those sub-zero climes of Labrador, at the same latitude on the other side of the Atlantic. By the time they arrive at high latitudes, these tropical waters have become dense and salty enough to sink to the ocean depths, carrying heat and dissolved gases such as oxygen and carbon dioxide with them into the abyss. These deep waters then travel vast distances over the ocean floor, eventually surfacing around a thousand years later in tropical waters, where the cycle begins again. The upwelling water carries nutrients from the ocean sediments, feeding photosynthesizing plankton at the ocean surface, which in turn remove carbon dioxide from the atmosphere, cooling the Earth. Here we catch a hint of how intimately water is connected via complex Gaian feedbacks to the biosphere and to greenhouse gases in the atmosphere, such as carbon dioxide, which in turn affects the distribution of temperatures on Earth's surface, determining which phase of water appears in a given place.

There is one final aspect of water that we must mention, even though we risk evoking a strong cultural taboo against such things. I refer to water's symbolic meanings. Symbols are images that point to ineffable, numinous, and hence only partially knowable qualities in the fabric of reality that bring tremendous healing for the psyche if rightly pondered and integrated into daily life. One such image is that of water as the elixir of life, as a powerful medicine for the soul, as a holy source of healing, appearing as such in countless myths and stories around the world to do with sacred fountains, wells, and springs. Temples and shrines are often located adjacent to such holy places. For the ancient Mesopotamians, the deepest waters represented the deepest wisdom. In Greek mythology, Aphrodite, the goddess of love, arose from sea foam created when Cronos threw the severed genitals of his father, Uranus, into the sea. In the Judeo-Christian tradition, water represents the spirit of God, the "fountain of living waters," the water of life, which, in the rite of baptism for Christians, removes sin. In the Hindu Vedas, water is experienced as deeply maternal, as an archetypally feminine element. In modern depth psychology, water represents the vastness of the unconscious psyche from which great insight and healing can emerge. Water can also take on more disturbing symbolic meanings, sometimes appearing as vastly dangerous turbulent seas racked by massive waves, as tsunamis, or as deep, still lakes full of menace. Water's light and dark symbolic manifestations reveal how the psyche, like Gaia, is a self-regulating system, compensating for one-sided attitudes with contrary images that guide the ego towards wholeness.

Could we gain a more profound understanding of the physical aspects of water if we contemplated these symbolic images gifted to us from the deepest layers of the psyche in conjunction with our scientific knowledge? I believe so, for I've noticed how I participate more fully with nature when I reconcile inner image and outer reality, when my awareness is transformed and enlivened by the integration of apparently irreconcilable opposites such as psyche and matter. A truly phenomenological exploration of water would therefore work in a highly personal way with these two mysterious aspects of water's ever-shifting presence – with its inner and outer appearances.

I return to the River Dart, to a stretch hardly touched by human hand. The river flows shallow here, and in the evening light a mosaic of pebbles and stones on the riverbed is bent into shifting shapes by the flowing water. Large trees extend moss-covered branches over the running water, confirming the river's wildness. A low-flying mallard duck splashes, feet splayed, into the dark water on the opposite bank. Water science and water symbol merge into an ineffable phenomenological experience as I witness all this. The river is no longer something exterior, something dead, alien. No. The river is alive, full of meaning and personality – a dream image, a non-human intelligence of tumbling water molecules who welcomes me into its living depths. The river opens me to the wholeness of water, to the wholeness of Gaia, and at last, I am home.

REFERENCES

Calvino, I. 1976. "Blood, Sea." In *t.zero*. New York: Mariner Books.

Harding, S.P. 2009. *Animate Earth: Science, Intuition and Gaia*. 2nd ed. Cambridge, UK: Green Books.

Harding, S.P., and L. Margulis. 2010. "Water Gaia: 3.5 Thousand Million Years of Wetness on Planet Earth." In *Gaia in Turmoil: Climate Change, Biodepletion, and Earth Ethics in an Age of Crisis*, edited by Eileen Crist and H. Bruce Rinke. Boston: MIT Press.

Lovelock, J.E.L. 1995. *The Ages of Gaia*. 2nd ed. Oxford: Oxford University Press.

McMenamin, M., and D. McMenamin. 1994. *Hypersea*. New York: Columbia University Press.

Weber, A. 2016. *Biopoetics*. New York: Springer.

2 Flow Motions and Kinethic Responsiveness

STEPHEN J. SMITH

A phenomenology of water experience can take as a starting point the bodily responses that oceans, as well as lakes, rivers, and streams, elicit from us. Having originated in the ocean and moved to land at some point in our evolutionary history, many of our motions must surely retain an affinity for the ocean even if they are now, for the most part, relegated to "fossil gestures" that can hardly be perceived (Hamilton-Paterson 2002, 159). Yet in the circulations of water, when bobbing on the waves, swimming in the surf, diving into the ocean depths, or taking in the ebbs and flows of water bodies from the shore, we can again sense the pulsing, throbbing, undulating motions of our earlier, water-immersed life forms.

Flow motions may well retain vestigial form in the most common yet compelling gestures of everyday life. The simplicity of waving to another may draw its animating power from the vitality that ocean waves express. The human touch may follow the sensuous line of action to the swell and trough of a wave building up to break. Even that most intimate of gestures, the kiss, may be felt as lapping and overlapping waves of wet emotion. Are these gestures of analogy? Are they merely motional projections of human emotion? Or are they indeed flow motions in which human affectivity is expressive of an original feeling for life and its inherent life forces? What further reminders are necessary that "oceanic memories continue among all humans who have landed – that the pulsing waves of ancestral amphibians are recorded in every undulation of an organ, in every sweep of tissue, in every course of blood" (Conrad-Da'Oud 1995, 311)? We need only to remember viscerally and impressionally those flow motions that have essentially "no inside, no outside, no up or down, no 'body' only wave motions, many kinds – short waves – long waves – dancing waves" (309).

In thinking about such flow motions and the bodily responsiveness they elicit, I want to consider how it is that the world can be approached ecologically, and in what David Abram (1996; 2010) termed a "more-than-human way." Such an approach would reveal not so much the world as a continuing backdrop to human actions, nor even a world beyond human affairs, but spaces of motile possibility and the corporeal capacities we have to move within them (Smith 2016). Particularly compelling in this regard are the forces of waves and the lure of oceanic waterscapes and the ways in which these flow motions and places of habitude become spoken for, acted upon, interacted with, or simply left to nature in what is probably the most telling gesture of respect. Imagine being on expansive surfing beaches mesmerized by the waves that change in shape and power throughout the day. Between the sitting and the surfing, there is the beckoning of the sea towards a more intense connection with the world. Think of being on the ocean, sailing or kayaking, lulled into a rhythm of cresting up and sliding down waves. Lifting, falling, lifting, falling, spray flies from the prow as the watercraft slices through the wave peaks. And think, too, right to the far reaches of human-scale consciousness, of the absolute terror hapless individuals must have felt when they saw water suctioned from lagoons and beaches and reappear as a towering tsunami. What dread these people must have felt in the face of forces that so mocked the human condition.

How separate are these waterscapes from human consciousness, from our perceptions of them, and from our possible actions in them? As the phenomenologist Maurice Merleau-Ponty (1962) wrote, "The relations between things or aspects of things having always our body as their vehicle, the whole of nature is the setting of our own life, or our interlocutor in a sort of dialogue. That is why in the last analysis we cannot conceive anything which is not perceived or perceptible" (320).

The water beckons to us. This invitation wells up in kinesthetic sensations of wading, plunging in, and diving under the waves. What may have appeared initially as having an independence and a separate material reality becomes a bodily responsiveness and motile correspondence. All about us in the folds and creases of the perceptual world, and noticeably in the shapes of waves, we can find ourselves moving fluidly with the elements. Which is not to say that, as with any relationship, things cannot go awry. Eddies, currents, whirlpools, vortices, spiralling columns of water possess powers that cannot be held back. There are forces beyond control for a "body animated with vortices and whirlwinds that are *sometimes* in phase with the movements of the world" (Serres 2010, 118; my emphasis). Ultimately, there can be no possessiveness at play here, except maybe the sense of being possessed. Even beckoning, soliciting, folding, enfolding, unfolding motions of a body of water are those to which we respond in bodily ways that lie inevitably beyond our human grasps.

There is a lessening of self-consciousness in this motile connection to the water. The purposeful, intentional consciousness that carries us through daily events, and in which we so often find ourselves swimming figuratively against the tide, or being overwhelmed and metaphorically carried out to sea, dissolves in the immediacy of actual contact with water. The flow motions experienced are not to be defined as acts of consciousness, but as openings to the world that fill us with impressions of the world's fluidity. There is a release from self-absorption and an openness to the animations of the world beyond the dry confines of anthropocentric existence. We recognize in such moments of release our fundamentally and "uniquely saturated presence" (Steinbock 1999) in the world which is reflected back in the form of ripples and waves of bodily sensation.

The present chapter is about becoming fluidly, vitally reminded and motilely, gesturally, ethically responsive to the primary movements of water. My initial focus is on the living experience of water flow, or flow motion, especially in recognizing the primacy of wave perception and the deep registers of wave motions in kinesthetic awareness. Gestures of life are then described as those that revitalize from moment to moment an interactivity with, and attunement to, the flow motions of the animate world. Being fomented and formed by movements that hold the relational dynamics of flow, we subsequently find there to be an "indisputable social determination of conduct," although it "is not something starkly superimposed from without" (cf. Casey 1998, 212). Indeed, there is, if not an "ethical imperative" (Lingis 1998), then a sense of corporeal "correspondences" (Ingold 2013) and "kinetic-kinesthetic-affective dynamics" (Sheets-Johnstone 2011) that are formative of more measured responses to waterscapes, their uses and preservations. I cast these actions and interactions that are attuned to water flow as motions of *kinethic* responsiveness.

Flow Motions

Flow, as a requisite feature of enjoyably sustaining movement (Csikszentmihalyi 1990; 1997; 2000), is the affective "chiasma" of consciousness that Merleau-Ponty spoke of in his posthumous work *The Visible and the Invisible* (1968). Here he described the reversible but never fully overlapping experience we have of being "flesh of the world" (139–55), simultaneously touched by the world while making bodily contact, seeing, hearing, and feeling ourselves in a folding, enfolding, unfolding bodily connection to the circulations, currents, rhythms, ebbs, and flows of the world. More than "an interlacing of a being with, potentially, all other beings," it is an overlapping fluidity, "an intersubjective plasma, as a complex of fluid elements that are just as capable of hardness as they are of liquidity" (Macke 2007, 401). Flow motion is not just my imprint on the things around me, but equally their effects on my own intentions. It is the kinesthetic and proprioceptive consciousness of moving while simultaneously being moved by forces and energies beyond myself. Drawing upon sensuous ecological renditions of Merleau-Ponty's idea of this flowing "flesh of the world" (e.g., Abram 1996; 2010), I will consider how certain flow motions that are essentially about being attuned to the world and receptive to the motions of a more natural order of living can be taken up as bodily responsive and ethical practices of water appreciation.

"Flow is of course what water excels in: Everything that flows is water; everything that flows participates in water's nature" (Macauley 2010, 46). A focus on waterscape gestures thus opens up a space for contemplating the flow motions of a human and more-than-human world. We look outwards from the constitution of self-identity through birth and heredity, from socialization and habitual practices, and from the effects of our emotional states. We peer right past self-identity to those relational gestures of mutuality and reciprocity that are about environmental attunement. We tap directly into a fluid manner of living responsively to the natural world (Smith 2006; 2007). Accordingly, we can think phenomenologically about flow motions and our bodily responsiveness to them and, instead of being confined to a denatured language of the body and abstract references to bodies of water, describe these flow motions as fully saturated, corporeal gestures. These gestures give telling indication of a range of natural expressivities intimately connected to moving, flowing, human and more-than-human life forms.

The wave stands out in this regard as a gestural exemplar. Ocean waves, rising as swells far out to sea, become sets of waves that crest and crash. Dumpers to body surfers or tubes that encase board riders, criss-crossing waves, signs of rips and undertows, bow waves and waves that form in the wakes of boats, wave after wave, waiting, waiting for the perfect wave. Monster waves originate in some tectonic earth movement in the ocean depths. Tidal waves, tsunamis, the cyclonic aftermath draw the adventurous to surfing nirvanas. Right down to waves that lap the beach, ripples in the shallows, shimmering waters where young children frolic. No two waves are alike. Each has a shape, a crest, a break that differs from the one before. Each wave has a chemistry, physics, biology, and ecology. Each wave has motion comprising circulations and sensibilities to which we can be perceptually attuned and physically connected. Each wave is a potential register of human interest, a sine of kinesthetic consciousness, a cipher of proprioception.

Unlike sound waves, heat waves, microwaves, and light waves that are sensed mostly in their effects rather than in their form, ocean waves are of an active human scale and, though "elemental, fluid, fickle, at once so tempting and yet utterly indifferent to the presence of man [sic]" (Grissim 2003, 54), they can be seen, kinesthetically felt, and caught in human gestures. They have wavelengths that register on scales of human movement. Of course, a wave can be ridden, body surfed, board surfed, wind surfed, or kite surfed. It can be sailed, kayaked, jumped over, and dived under. But a wave can also be caught in more subtle actions. In the correspondence mentioned already between ocean waves and hand waving, the former primarily beckons towards the sea and the sensuality of interaction while the latter both beckons towards and signals a departure from interactions with others who have a more clearly defined and demarcated sensibility. The wave is part of a flow experience that is both worldly and deeply interactive. And just as there is potential danger in underestimating the power of ocean waves, the gesture of waving cannot always be contained – one has to go with the flow and ride the wave. A wave of recognition that is contained feels wrong. Either it is a wave that is not really called for or it is inadequate in its recognition and appreciation of the person seen. The wave is a primordial gesture of being with others and the otherness of the world. Waves lap and crash and rock and lull. They roll and pummel and caress and kiss. The distant oceanic wave approaches, like the wave of someone I begin to recognize, as a motion that invites action and human interaction. Dare I wave back? Can I do otherwise?

Clarice Lispector (1974) wrote, "There it is, the sea, the most incomprehensible of non-human existences. And here is the woman, standing on the beach, the most incomprehensible of living beings ... Their mysteries can only meet if one surrenders to the other: the surrender of two unknowable worlds, made with the confidence with which two understandings give themselves up" (162). Surfers know this surrender well, referring, for instance, to how "[w]omen surfers tend to go with the flow rather than battling the break ... women surf with such fluid grace because there is a unique symbiotic relationship between them and the water" (Dixon 2001, 17). Roger Housden (1993) expressed a similar feeling of surrender when he wrote: "Somewhere in the heart of the crashing wave, at the heart of the listening human, is a sound so delicate and constant it is almost unbearable to hear. It is utmost ecstasy, and beckons our dissolution" (180).

Waves, as quintessential flow motions, inevitably peter out, disperse, exhaust themselves, or, shall we say, evaporate. Their chiasmic fluidities are drawn back into oceans, lakes, and reservoirs, while also releasing vapours and humidity that coalesce as clouds of condensation. I want to ask now about the gestures of life that flow with this hydrology. What becomes atmospheric about vapourized flow motions? What then falls to the earth? And what, in effect, and especially in affect, is precipitated when sensing a cycle of water that is no longer simply the backdrop to human affairs?

Gestures of Life

Bertrand Russell (1996) posed the question, "What is the difference between an event that is not experienced and one that is?" He went on to say that "[r]ain seen or felt to be falling is experienced, but rain falling in the desert where there is no living thing is not experienced. Thus we arrive at our first point: there is no experience except where there is life. But experience is not coextensive with life" (725). Yet, where there is rain there is, if not life, then surely its possibility, and if not experience, then surely there are animations constitutive of living things. A rain *event* is, in so many experiential senses, testament to life, yet larger than any particular life experiencing it. It is the event of bestowing, generating, and regenerating life in a multitude of ways that surpasses what we can come to understand through particularly situated experiences.

There is a vital generalization to the event of rain, which is to say an eventfulness that registers sentiently as well as nutritively. The desert comes to life. Scorpion weed, prickly leaf, brittlebush, and Indian paintbrush bloom while raptors locate newly emergent terrestrial prey. The event itself inevitably and eventually passes consciously and summarily, which is to say anthropocentrically, into a record of millimetre rainfall counts, runoff rates, and reservoir levels. But in the first instance there are simply moments of animation. Before extrapolating and appropriating the event, and within what Martin Heidegger called the "enowning" that particularizes an event of rain as that which is experienced in its situated, existential meaning (see, e.g., Polt's 2015 account of *Ereignis* in Heidegger's writings), there are stirrings, flutters, ripples, waves of emotion, circulations of desire, intensities of affect. Flow motions are most evident impressionally in multiply nuanced expressions of a particular rain event.

We experience gathering clouds in times of severe drought with raised hopes that teasingly, maddeningly, dissipate with the cloud scatterings. Days, months – years, even – feel like a never-ending ebbing tide of drained emotion. Yet a darkening sky, freshening winds, and thunderclaps heard in the distance lift human hearts in joyous anticipation. A wave of revitalizing, rejuvenating, regenerating emotion builds. The first drops come hissing and spitting into the arid dust. Staccato sounds at first, they quickly crescendo to a torrential downpour. Clouds burst. Sheets of water fall. The rain soaks the skin of the parched earth, seeping deep into the subterranean layers. Runoff pours into trenches and canals, engorging waterways to overflowing.

Monsoons, tropical cyclones, and hurricanes: these storm events of noise and fury drown out the rain sensations of breaking drought. Deluging the senses, such storms precipitate water floods, flushing the earth in torrents of effluence. Rainfall, however, in breaking droughts and in bursts and rushes of feeling that accompany the life-giving, life-renewing events of rain, is felt as "a life gesture" (Bachelard 1971, 122) and expressed accordingly. Like the boon-granting hand gestures of Buddhists, drought-breaking, life-giving rain is greeted with an open-armed embrace. We raise our arms as the rain falls, feeling, from the initial drops to the cloud-bursting downpours, the "vitality affects" (Stern 1993; 2004; 2010) of a saturating joy. Heaven and earth are felt from head to toe and held within the span of one's arms.

Falling rain, precipitation, belies, in the case of drought-breaking rain, the etymology of an evil from heaven, angels cast out, falling down to earth precipitously. On the contrary, there is energy, exuberance, playfulness even, and moments, passages, durations of uplifting joyousness. "*Eureka*! I've felt, he says, the force of the wave that carries my body. But what second force makes it gush, now and in addition, out of the water?" (Serres 2010, 64–5). A forceful gushing, a seemingly alchemical attraction to the falling rain, makes for dancing with joy, singing in the rain, collecting, channelling the downfall through flows of affect that remain effectively on a scale of emotional satisfaction and bodily satiation. Fluidity of movement is an interaction affect comprised from moment to moment of bursts and gushes, rushes, flows, ebbs, and flushes. It is, more than an attraction to water; it is a way of taking on of the very qualities of water flow and rainfall events. Well might we speak of not just interactions but, rather, motile "correspondences" (Ingold 2000, 199; 2013, 105–8) between the climatic acts of water falling and the biotic acts of joyful responsiveness.

Richard Rojcewicz (2014) argues for piety rather than hubris in taking up Thales's comment that "everything is water." Rojcewicz writes, "Respect for water, for example, as a precious gift, leads to a very different practice than ravaging – to conservation certainly but, more than that, to something like cooperation with the things and forces of nature rather than opposition to them" (220–1). He warns of the abstractions of water, as "H_2O," and yet something dangerously abstract sounds in the very notion of water, as if it is "given to us from a mysterious source" rather than in the waves, rains, and flows of sensed events. If Thales was right that "water is everything," then it is so as an interactional affect of the gestures of life that are wetted, watered, given wavelengths, and flow literally between heaven and earth. Water may well seem at times "a gift from the gods" (221), yet the gift is not water per se but the life and lives animated by rainfall events.

But rivers and streams flow back to the ocean. Reservoir levels fall. And the porosity of the earth is pressed into the service of rising water demands. What gestures of life make sense of this inevitable scarcity? What human responsiveness to water insecurity remains in keeping with the flow motions of life animation?

A Thirst for Life

An often-cited line from Gaston Bachelard's *Poetics of Reverie* (and one he attributes originally to Franz von Baader) is that "[t]he only possible proof of the existence of water, the most convincing and the most intimately true proof, is thirst" (Bachelard 1971, 178). The thirst Bachelard surely had in mind was not necessarily of a momentarily quenchable, slakable kind, since it was a greater desire than the cravings, longings, yearnings for anything less life essential. Even a thirst for knowledge pales in relation to that first and essential thirst that proves not just the existence of water but the existence of life itself.

> Suddenly the first Thirst comes glistening out of the water, the first Adoration comes back, such a thirst was only known in paradise, coming out dripping wet, Passion recovered, a Thirst exciting as love, holding up palms full of water; in the tale when one bows, it is, all of a sudden, the sleepiness which awakens, the submerged which arises, the palms are full of meaning, the secret of water was forgotten, but when one wishes for it with all of one's being, one loves it anew as if it was loved at the beginning. (Hélène Cixous, cited in Fisher 1999, 114)

Thirst-quenching waters invigorate and energize. They restore the primordial feelings of being alive. Perhaps we need to speak less of slaking and quenching thirsts, as if mollifying an unpleasant state, and more of realizing a repeatable pleasure in being alive.

Well water, spring water, and tap water require minimal conditions for safe consumption. But to drink in a thirst-quenching way is a fundamental gesture of life affirmation that releases us not into intestinal safety so much as into the full-bodied satisfaction of satiated thirst. Dry, parched throats are salved in the saving grace of a cool drink of water. Yet this experience of water is not shared by everyone. In places where drinking water is in dire scarcity, "[n]o longer do all people have the right to quench their thirst; this is a right held exclusively by the rich" (Shiva 2002, 102). There are, moreover, unquenchable thirsts and a price put on thirst where corporations bank on it through the commodification and privatization of water systems (Piper 2014). Drinking water is levied, apportioned, bottled, and sold. Unquenchable thirsts become a reality for the increasing numbers of people living without water ready to hand and for those bloated economically with water surpluses.

What will the ethical response be to this widening disparity? Surely it requires much more than addressing the unevenness, the imbalance, of water interests through enhanced provisions of water security for the thirsty peoples of the world. Surely it must be more than asserting a human right, taking into account the quality of the water to which people are entitled as not merely available in sufficient quantity, safely drinkable, and physically and economically accessible (cf. Doorn 2013, 103), but as essentially thirst-quenching. Surely all peoples have, if not the legal right, then the birthright to quench their thirsts in the pleasure of drinking life anew.

Ethical responsiveness arises with and from the gestures of life. Giving drink to another whose thirst exceeds my own is a fundamental, if not *the* fundamentally ethical gesture of life affirmation. It is an action springing from "the original know-how of life" (Henry 2012, 97), which already knows its proper sustenance. Indeed, as Michel Henry goes on to say, "Every possible ethics is rooted in the immanent teleology of life. This is true not only for the theoretical or normative ethics that represents ends or values, but also for the original ethics, or rather *ethos* itself, that is the set of continually renewed processes in which life carries out its essence" (97).

A vital continuity is thus intuited between flow motions and the gestures of ethical life. Suffering from thirst creates the bodily imperative to obtain water, whether by one's own efforts or from the hands of others. For in the parchedness lies a suffering that is all the more compelling for its joyful conversion through drinking water that is freely offered. Life, as Henry repeatedly insisted, vacillates between suffering and joy. The former is the precondition for the latter (37). And in this pathic subjectivity, this auto-affectivity that both suffers and rejoices, there can come an alignment of affective life and effective social, ecological, and political sensibilities. Donating water to thirsty persons is the gestural form of the auto-donation of life to its self-augmentation. This is the "I can" of a praxis that reaches out to world concerns of water scarcity from the vantage point of the simplest life-affirming gestures. It is an

"I can" in which "world-content and the execution of our potentialities are inextricably bound together" (Staudigl 2012, 348).

Water Ethics

Have I created a distracting water imaginary – a water reverie, if you will (cf. Bachelard 1983) – in the midst of pressing matters of dammed waterways, the depletion of aquifers, pollution of water reservoirs, and imminent water wars? Has this phenomenological rendering of water gestures, waves and flows, rainfall, and runoff left unaddressed the most pressing questions of life? Certainly Michel Henry's (2015) radical phenomenology of the flesh of auto-affectivity and intimations of world connectedness urge taking indications of water gestures to heart and to considerations of living between water plenty (rain) and poverty (thirst) – which is to say, to the fullness of life within the span of joy and suffering. But might the life-worlds of water then be something more than just the dispositional and transpositional state of mind prompted by this life phenomenology?

The shifting foci of environmental concern, which I understand to be primarily about recognizing the perversion of the natural dynamics of flow motion, seem bathed in a phenomeno-logic of shared corporealized sensibility of, and for, waterscapes and for the sake of their healthy and sustainable animations. So, let us not merely imagine but become deeply immersed in the movement flows of the *Blue Covenant* (Barlow 2007) that bring to life nothing less than an essential affinity for water. The point is not just to contemplate such flows, but to create practices of environmental animation that ride upon and dive into, receive with open arms, bathe in, and drink that most elemental source of life on the planet. This is no mere diversionary dalliance, nor simply a laudable commitment to sustainable living and to the possibilities of human flourishing, but fundamentally it is a felt imperative springing from our bodily, carnal, corporeal connectedness to that which is life itself or the closest we shall come, in human terms, to realizing what Henry (2008) referred to as the "auto-affectivity" of life, as the "essence of manifestation" of all worldly events. What is at stake is not so much even the movement of phenomenological thought about water experience, but movement itself, flow motions, which can be felt in their "own being, only when the world has lost its power" as an external, thought-about reality (Henry 2009, 43).

A phenomenological sensitivity to, and practical incorporation of, flow motions may well afford appreciation of the actions, practices, commitments, and accords now needed to fully address the facts of shifting and worrying hydrological cycles. For just as waves, evaporation, rain, and thirst indicate a cycling, animating circle of life-sustaining gestures, the gestures they prompt show that the flow of water is not simply subject to natural cycles of evaporation, condensation, precipitation, absorption, and runoff. Flow motions are constitutive of a "hydrosociality," according to which "water is conceived not as a self-identical object but as a process whose identity is formed in social relations" (Linton 2010, 224). More than an intersection of "the human economy and the earth's hydrological system" (Simmons, Woog, and Dimitrov 2007), the "hydrosocial cycle" is "a process in which flows of water reflect human affairs and human affairs are enlivened by water" (Barlow 2013). Indeed, "every instance of water is secondary to the process of engagement that makes it part of our world" (Linton 2010, 224). "Apart from drinking it, our physical engagements with water can have the potential to change things" (236). Practices of water conservation and sustainable management literally flow from the iterative experiences we have with water bodies, their forms, patterns, cycles, rhythms, and gestural affects.

A phenomenologized hydrosociality, which is to say a socially situated, ecologically referenced, and inherent sensitivity to flow motions, gives particular emphasis to water ethics. It is recognized, for instance, that

> water management must expand to include an ethic of virtue – where the quest is not only for improved decision-making frameworks, but is also for persons who act out of recognition that scientific and technological knowledge must be situated in relation to deliberations on ethics, fairness, temperance, and justice and which includes great humility regarding the types and ends of human knowledge. In this sense, a water ethic must be seen as moderating, rather than managing, the human-environment relationship. (Brown and Schmidt 2010, 280)

Yet bodily responsiveness to flow motions promises more than "an ethic of virtue" and more than a means of "moderating" the "human-environment relationship." Water gestures, as I have described them in terms of responsiveness to flow motions of the hydrosocial cycle, afford life-sustaining, life-affirming indices of the best practices of water management and conservation.

Water access and availability are considerations consistent with the enlivening movements of bodies of water. Currents, eddies, vortices, and even tsunamis are movements that test the best-laid management plans and practices, as do evaporations, precipitations, permeations, and run-offs. Water security in strictly scientific and technological terms is ultimately an impossibility, given the uncontainable movements and axiomatic fluidity of water. As the Zen philosopher Alan Watts once remarked,

> But you cannot understand life and its mysteries as long as you try to grasp it. Indeed, you cannot grasp it, just as you cannot walk off with a river in a bucket. If you try to capture running water in a bucket, it is clear that you do not understand it and that you will always be disappointed, for in the bucket the water does not run. To "have" running water you must let go of it and let it run. (1968, n.p.)

Water ethicality attuned to life involves more than meeting life's minimal biological preservation. It means being released to life's animations, sufferings, and joys. It means letting go our grip on life and going with the flows of water. Kinethic responsiveness to water's flow motions is, in effect, comprised of "gestures of ethical life" (Kleinberg-Levin 2005) that pulse, spring up, wash over us, rain upon and fill us, and seep into our lived and living relations with one another.

Conclusion

Flow motions and water-responsive gestures intimate, in essentially human terms, what is happening, and can happen, in larger-scale environments – rivers, estuaries, oceans – for other animate beings. These motions afford somatic points of departure for seeing, feeling, imagining the animations of places that can sustain life well beyond what any of us really know about it but in which, in the responsiveness of gestural life, we may find the expression of an auto-affectivity and auto-donation of life itself and the motive for ensuring our own and other life forms may flourish.

A line of flow is thus followed from elemental flow motions to life-receiving gestures to gestures of ethical life. The "kinetic-kinesthetic-affective dynamics" (Sheets-Johnstone 2011) of ocean waves, joyous gestures of rainfall, as well as the suffering gestures of thirst can be cast as a proto-ethicality, a corporeal ethics – which is to say, as kinethic responsiveness to, and vital affectivity for, wider circulations, cycles, ebbs and flows, waves and turbulences, stagnations and vortices of bodies of water. The line of flow can be extended still further to an explicitly ethical responsiveness to water abundance as well as scarcity, to saturation and to thirst, and even to a stewardship of those "bodies of water as eyes of the earth" (Macauley 2010, 297) that are less healthy than they ought to be for all thirsty, suffering life forms.

Undulating oceans, falling rain, flowing streams, and healthy river systems not only comprise a hydrological cycle, they disclose a hydrosociality that manifests the dynamics of this far-reaching, ethical responsiveness. Water circulates gesturally. Abundance and scarcity are experienced in movements of joy and generosity. An eco-phenomenology of water affinity may thereby come to seep into, permeate, infiltrate, indeed flow into fuller civic immersion in pressing matters of water policy and politics. What is at stake is nothing less than an ever-deepening and ever-widening responsiveness to the flow motions that sustain life in its manifold vitality.

REFERENCES

Abram, D. 1996. *The Spell of the Sensuous: Perception and Language in a More-than-human World*. New York: Vintage Books.
– 2010. *Becoming Animal: An Earthly Cosmology*. New York: Pantheon Books.
Bachelard, G. 1971. *The Poetics of Reverie: Childhood, Language, and the Cosmos*. Translated by D. Russell. Boston: Beacon Press.
– 1983. *Water and Dreams: An Essay on the Imagination of Matter*. Translated by E.R. Farrell. Dallas, TX: Dallas Institute Publications.
Barlow, M. 2007. *Blue Covenant: The Global Water Crisis and the Coming Battle for the Right to Water*. New York: W.W. Norton.
– 2013. *Blue Future: Protecting Water for People and the Planet Forever*. Toronto: House of Anansi Press.
Brown, P.G., and J.J. Schmidt. 2010. "An Ethic of Compassionate Retreat." *Water Ethics: Foundational Readings for Students and Professionals*. Washington, DC: Island Press, 265–86.
Casey, E. 1998. "The Ghost of Embodiment: On Bodily Habitudes and Schemata." In *Body and Flesh: A Philosophical Reader*, edited by D. Welton. Oxford: Blackwell, 207–25.
Conrad-Da'Oud, E. 1995. "Life on Land." In *Bone, Breath, & Gesture: Practices of Embodiment*, edited by D.H. Johnson. Berkeley, MA: North Atlantic Books, 297–312.
Conrad, E. 2007. *Life on Land: The Story of Continuum, the World-renowned Self-discovery and Movement Method*. Berkeley, MA: North Atlantic Books.
Csikszentmihalyi, M. 1990. *Flow: The Psychology of Optimal Experience*. New York: Harper and Row.
– 1997. *Finding Flow: The Psychology of Engagement with Everyday Life*. New York: Basic Books.
– 2000. *Beyond Boredom and Anxiety: Experiencing Flow in Work and Play*. San Francisco: Jossey-Bass.
Dixon, P. 2001. *The Complete Guide to Surfing*. Guilfort, CT: The Lyons Press.
Doorn, N. 2013. "Water and Justice: Towards an Ethics of Water Governance," *Public Reason* 5 no.1: 97–114. https://ethicsandtechnology.eu/wp-content

/uploads/2013/04/Doorn2013_PR_Water-and-justice-towards-an-ethics-of
-water-governance.pdf.

Fisher, C.D. 1999. "Cixous' Concept of 'Brushing' as a Gift." In *Hélène Cixous: Critical Impressions*, edited by L.A. Jacobus and R. Barraca. Amsterdam: Gordon and Breach, 111–22.

Grissim, J. 2003. "California: Fever and the First Wave." In *Big Wave: Stories of Riding the World's Wildest Water*, edited by C. Willis. New York: Thunder's Mouth Press, 49–62.

Hamilton-Paterson, J. 2002. "Sensing the Oblique." In *Down Time: Great Divers on Diving*, edited by E. Kittrell, C. Kittrell, and J. Kittrell. Austin: Look Away Books, 153–65.

Henry, M. 2008. *Material Phenomenology*. Translated by S. Davidson. New York: Fordham University Press.

– 2009. *Seeing the Invisible: On Kandinsky*. Translated by S. Davidson. New York: Continuum.

– 2012. *Barbarism*. Translated by S. Davidson. New York: Continuum.

– 2015. *Incarnation: A Philosophy of Flesh*. Translated by K. Hefty. Evanston, IL: Northwestern University Press.

Housden, R. 1993. *Soul and Sensuality: Returning the Erotic to Everyday Life*. London: Random House.

Ingold, T. 2000. *The Perception of the Environment: Essays on Livelihood, Dwelling and Skill*. New York: Routledge.

– 2013. *Making: Anthropology, Archeology, Art and Architecture*. New York: Routledge.

Kleinberg-Levin, D.M. 2005. *Gestures of Ethical Life: Reading Hölderlin's Question of Measure after Heidegger*. Stanford, CA: Stanford University Press.

Lingis, A. 1998. *The Imperative*. Bloomington, IN: Indiana University Press.

Linton, J. 2010. *What Is Water? The History of a Modern Abstraction*. Vancouver: UBC Press.

Lispector, C. 1974. *Soulstorm: Stories by Clarice Lispector*. Translated by A. Levitin. New York: New Directions.

Macauley, D. 2010. *Elemental Philosophy: Earth, Air, Fire, and Water as Elemental Ideas*. Albany, NY: SUNY Press.

Macke, F.J. 2007. "Body, Liquidity, and Flesh: Bachelard, Merleau-Ponty, and the Elements of Interpersonal Communication." *Philosophy Today* 51 no. 4: 401–15. https://doi.org/10.5840/philtoday200751425.

Merleau-Ponty, M. 1962. *Phenomenology of Perception*. London: Routledge and Kegan.

– 1968. *The Visible and the Invisible*. Evanston: Northwestern University Press.

Piper, K. 2014. *The Price of Thirst: Global Water Inequality and the Coming Chaos*. Minneapolis: University of Minnesota Press.

Polt, R. 2015. "The Untranslatable Word? Reflections on *Ereignis*." *Journal of Chinese Philosophy* 41 no. 3–4: 407–25. https://doi.org/10.1111/1540-6253 .12112.

Rojcewicz, R. 2014. "Everything Is Water." *Research in Phenomenology* 44 no. 2: 194–211, https://doi.org/10.1163/15691640-12341285.

Russell, B. 1996. *History of Western Philosophy*. New York: Routledge.

Serres, M. 2010. *Biogea*. Translated by R. Burks. Minneapolis: Univocal.

Sheets-Johnstone, M. 2011. *The Primacy of Movement* (expanded 2nd ed.). Philadelphia: John Benjamins.

Shiva, V. 2002. *Water Wars: Privatization, Pollution and Profit*. London: Pluto Press.

Simmons, B., R. Woog, and V. Dimitrov. 2007. "Living on the Edge: A Complexity-informed Exploration of the Human-Water Relationship." *World Futures* 63 no. 3–4: 275–85, https://doi.org/10.1080/02604020601174927.

Smith, S.J. 2006. "Gesture, Landscape and Embrace: A Phenomenological Analysis of Elemental Motions," *Indo-Pacific Journal of Phenomenology* 6 no. 1: 1–10. https://doi.org/10.1080/20797222.2006.11433914.

– 2007. "The First Rush of Movement: A Phenomenological Preface to Movement Education." *Phenomenology & Practice* 1 no. 1: 47–75. https://doi .org/10.29173/pandpr19805.

– 2016. "Movement and Place." In *Encyclopedia of Educational Theory and Philosophy*, edited by T. Saevi. New York: Springer.

Staudigl, M. 2012. "From the 'Metaphysics of the Individual' to the Critique of Society: On the Practical Significance of Michel Henry's Phenomenology of Life," *Continental Philosophy Review* 45 no. 3: 339–61. https://doi.org/10.1007 /s11007-012-9226-9.

Steinbock, A.J. 1999. "Saturated Intentionality." In *The Body: Classic and Contemporary Readings*, edited by D. Welton. Malden, MA: Blackwell, 178–99.

Stern, D.N. 1993. "The Role of Feelings for an Interpersonal Self." In *The Perceived Self: Ecological and Interpersonal Sources of Self-knowledge*, edited by U. Neisser. Cambridge: Cambridge University Press.

– 2004. *The Present Moment in Psychotherapy and Everyday Life*. New York: W.W. Norton, 23–54.

– 2010. *Forms of Vitality: Exploring the Dynamic Experience in Psychology, the Arts, Psychotherapy, and Development*. Oxford: Oxford University Press.

Watts, A.W. 1968. *The Wisdom of Insecurity: A Message for an Age of Anxiety*. New York: Vintage eBooks.

3 Creaturely Migrations on a Breathing Planet

DAVID ABRAM

A Flock of Selves

The desert drips silence. A few sounds are evident – a mouse skittering through dry leaves, the lapping of breeze-stirred ripples at the edge of this flooded field – yet all within the compass of a silence that radiates from the ground and leaks from the sagebrush and the bare twigs of the cottonwoods clustered on the far side of the water. Whispers of cloud laze in the stillness above the western hills, and a giant sun slides down between those wisps, stopping to rest itself on the skyline, flattening and widening as it does so. The sun seems reluctant to leave this domain, pouring its honeyed gaze upon the clumps of sage and the cottonwood branches, granting a gold sweetness like that of apricots to the huge, implacable quiet. The sun slips its toes behind the hills, then sinks down into the earth.

We are at the northern edge of the Chihuahuan Desert, in late winter, standing in the *bosque*, the sparsely wooded flood plain of the Rio Grande River. Near this strip of moist bottomland, a tribe of Piro Indians built their pueblos some seven hundred years ago; they hunted here, and gathered wild fruit, and raised turkeys until Apache raids and European diseases forced them to abandon these fertile soils. Later, the Apaches themselves came to camp among the willow stands and the cottonwoods. In recent years, the river has dwindled (its waters drawn for use by industries farther north), yet now and then it still floods these riparian fields, enabling the local farmers to grow alfalfa and corn in the desert silence. *Acequias* – irrigation ditches – run alongside the fields. Locals use these to manage the water levels not only to benefit their crops, but on behalf of the many wild migratory birds that gather in this narrow place of moisture every winter, a seasonal convergence much older than even the earliest human inhabitants.

At the moment, only a single pair of ducks – common mergansers – paddles along the near surface of the flooded plain. To my left, the cloud wisps are bright pink brushstrokes above the hills. They deepen slowly to magenta as I watch. When I turn back towards the east, a band of dark grey is rising from the horizon: the visible shadow cast by the Earth into its atmosphere.

Cool air wafts against my face, a chill coming up from the water. A duck out near the middle is speaking gently: three staccato quacks, then three more. And then I hear this other noise: the faint sound of a screen door with rusty hinges slowly opening off to the north. Some old-timer's cabin? But then another rusty door opening, and then another, and then three and five more – until suddenly it seems a thousand rusty hinges are complaining louder and louder as they approach out of the dusk, becoming visible now as a silhouetted arc of birds, huge, flapping towards us from the north, a wobbly configuration stretching a half-mile or more across the sky, getting larger and larger as the crazed, rusty bugling grows in volume. And then I notice another arc darkening behind the first one, and then a series of wobbly lines converging from the northeast, and others from the northwest, as the swelling, creaky-hinged crowd of voices begins to vibrate within my chest, an immense and enveloping sky-sound of the most shivering weirdness, a rapid "ah-uh-uh-uh-uh!" echoing from and overlapping through every sector of the atmosphere, as thousands upon thousands of sandhill cranes descend upon the flooded field before me, some of them swerving in just above my stunned and upturned face, their wingtip feathers splayed, others circling and circling before they land in small and large sprays of water upon the now cacaphonous surface, while other clusters and lines steadily emerge out of the north, the sky itself nothing other than a shifting, breathing fabric of wings and outstretched bugling necks.

Their croaking cries reverberate in the hidden hollows of my body and I'm carried out of myself by waves of intersecting rhythm, my awareness scattered and radiating across the water into the trees, vibrating off behind me into the clumps of sage and echoing out from my shoulders into the deepening dusk. I am dispersed; without an inside or outside, I am a sheer multiplicity, a crowd, a flock of selves, or cells, echoing one another across the fields – less a body than a fluid medium rippling with interpenetrating rhythms – a shifting focus whirling from the hills to the trees to the tears streaming down my partner's cheeks as she, too, is carried aloft by the cranes.

These same pterodactylic creatures gathering here in immense, sociable crowds to roost and to feed in the stubbled fields (a few adults standing extra tall, their necks extended to keep watch for the whole flock) were fiercely *antisocial* a few months earlier in their breeding grounds 1,600 kilometres north. Last summer, each mated pair (sandhill cranes mate for life) defended an expansive territory just for themselves and one or two newly hatched colts, whom they steadily fed on insects, seeds, and small rodents. Yet as the early autumn cold began to clamp down on the insects and other foods in those mountains, these crane families began feeding and roosting with other nearby families. The young cranes made more and more practice flights, impelled by other scattered groups whose coarse bugling they heard gliding far overhead until, unable to resist, they and their parents took to the sky, rising and rejoining others in strands that gradually gather into great, spiralling hordes, fractalled dragons made of hundreds of long-necked and exuberant voices, soaring and flapping towards the south.

How does it feel to add oneself, after a summer's solitude, to this slowly gathering sky-torrent? By what landmarks do these aggregate dragons navigate? How does this flood of wings find its way every autumn, year after year since time immemorial, to this tiny oasis of moist bottomland in the immensity of the New Mexico desert?

Other sandhill collectives spend these cool months far to the east and the south, in Texas and northern Mexico. Come late winter, waves of fidgety restlessness will spread through their communal roosts, and the young males in those flocks will find themselves leaping more concertedly into the air, the moment ripening within them and their fellow leapers until they all ascend in whirling commotion, making their way towards their remembered staging ground along a small stretch of the Platte River, in Nebraska. Here they'll convene for a few weeks with *half a million* of their tribe – four-fifths of the world's sandhill cranes – feeding and fattening themselves in the cornfields, strengthening themselves for the longer part of the journey. Then this broad, feathered blanket of birds will flap skyward and begin to unravel into its different threads as families separate out from the fabric and fly towards their respective breeding grounds in Saskatchewan and Alaska and Siberia, each bonded pair with its one or two offspring following the remembered route towards the individual nesting site in the boggy tundra where they'd nested last year, and the year before, and the year before that ...

What is this allurement, this seasonal memory that rises in the muscles, calling one skyward, drawing one back and back to the place of one's begetting, to that precise blend of wind and rock and glistening water? The irresistible draw of the bustling and clamorous crowd now giving way, as it always has, to the imperatives of solitude and intimacy and home?

Fluttering Conundrums and Finned Imperatives

Unlike other butterflies, monarchs undertake a long annual migration similar to that of the cranes and many other birds. Those monarch butterflies who emerge from their chrysalis stage in eastern North America to feed on the nectar of milkweed and golden rod and clover launch themselves on a mass southward migration covering as many as 5,000 kilometres to reach their overwintering sites on great conifer trees in the mountains of south-central Mexico. Filigreed with black, their gold wings flutter and dip across yards and fields, beguiling us with bright hues that signal their toxicity to potential predators. When they arrive on the transvolcanic plateau of Mexico, they'll cluster in immense numbers on the branches and trunks of the Oyamel fir trees, their metabolism slowing down as a quiescence settles over the flocks. Those that survive the winter hibernation will rouse themselves in late February to mate and begin the long migration north. But they'll fly only as far as they need to reach the first stands of milkweed in the southernmost states. Then they lay their eggs on those milkweeds and die. The caterpillars that hatch from those eggs a few days later will feed on the milkweed for a couple of weeks before transforming themselves into pupae. When butterflies emerge from those chrysalises, they'll continue the journey north for several weeks until the females, once again, lay their eggs on stands of milkweed and pass away. Only the *third generation* born of that northward migration will reach its ancestral destination before laying its eggs. And the subsequent generation – the one born in that northernmost homeland – will not die after four or five weeks, like its parents and grandparents and great-grandparents. The butterflies of this special generation will live seven or eight *months*; it's they who undertake the whole southward migration, travelling upwards of 80 or 95 kilometres a day. *Four generations removed from those who last journeyed south,* they will wing their way over mountains and spreading suburbs and dammed-up rivers, roosting in maples or pines or willows only to push south afresh the next morning, ultimately zeroing in on the very same few acres of conifers in the Mexican highlands – perhaps even the very same tree – to cluster with a hundred million other monarchs through the winter. If they survive, it is they who'll mate when the weather begins to warm, and begin again the spring migration back north.

Unlike cranes and other migrating birds, then, these butterflies can't possibly rely on the remembered landscape for navigation; the monarchs flapping their way south and finally converging on those precise trees are the great-great-grandchildren of those who made the same journey last year! By what magic do these delicate insects find their way? How does an organism inherit such intricate instructions – precise navigational guidance that must be different for each successive generation? We do not know. Recent research suggests that the monarchs may discern their direction by the position of the sun, using their internal circadian rhythms to vary the angle at which they orient themselves throughout the day. Still, a monarch flapping southward through Georgia would need to interpret the angle very differently than a migrant passing at the same latitude through Mississippi or Texas. When high winds or storms blow the butterflies far off course, how do they recalibrate their orientation? During the spring migration, even along a single route north, a butterfly from the first generation would need to orient by the sun very differently from a butterfly of the second or third spring generation. We have no idea how any creature, much less such a diminutive insect, could pull off such a thing.

We had beached our kayaks on one of the larger islands for the night. After a simple meal, I went walking along the coast as the sun slipped down towards the horizon, drinking the salt air and listening to the lapping of the small waves and the wind in the needles. After some time, I came to the edge of a surging stream about 3.5 metres across, whose surface was rippling and splashing in the fading light – and without paying much attention, I sat down on a mossy rock a ways back from the stream's edge just to bask in the rushing speech of those waters and to gaze out into the oncoming night. And I lost myself in some reverie or other, until my awareness was brought back to the place by a pale glow beginning to spread into the sky from the rocks on the far side of the stream. The glow got steadily more intense until, as I watched, the full moon was hatched from those rocks, huge and round as a ripe peach, pouring its radiance across the stony beach and the gleaming waves and the rustling spruce needles and generally casting a kind of spell over the whole place.

Now, I have never, of course, seen a cow jump over the moon. But that night I did see a fish jump over the moon. A great streamlined silhouette, its tail flapping, arced right over the full moon and splashed back into the water. I couldn't believe what I'd just seen, and so was still staring at the after-image – and then another silhouette leapt right over the moon!

I got up and walked over to the water's edge: the stream was thick with salmon, boiling with salmon, all jostling and surging against the current in fits and starts – it was as if the stream were made of salmon! I gazed and gazed for a couple of hours. Then I went back to my tent and tried to sleep, but couldn't. So I came back in the middle of the night and stood staring into that moon-illuminated river of fish, and then I waded out into the middle of that mass of silvery, sparkling muscles all surging and lunging against the current. In the middle of the stream, I was up to my knees in salmon, but they didn't care – didn't even notice; they bumped into my legs and then plunged on past with a single-minded determination I'd never encountered before, swimming past their dead or dying siblings and cousins as they floated back downstream on their sides, with their mottled skin beginning to fall off. The earnest salmon around me just nudged them aside, hardly noticing, intent on one thing and one thing alone – getting upstream to their spawning place, depositing their eggs and fertilizing them before they too began to fall apart and to die. I'd never imagined such intensity. Their total focus on getting upstream to create new life, and their utter obliviousness to everything else – to their dead or dying relatives, to other species that might prey upon them (to me, for instance, my own legs shivering among them), to everything other than the impulse to procreate ... I'd never before encountered such a collective visceral imperative.

When I returned to the splashing stream the next morning, an earlier high tide had erased most of my night-time footprints from the sandy streambank. The waters, however, had not erased the massive, clawed prints of a large grizzly. Several half-eaten salmon carcasses lay about on the rocks. I tracked the bear up into the spruce forest and part-way around a boggy meadow; then my jitters got the better of me and I hiked back to my camp.

Ocean-Going Avatars of Place

Pacific salmon are anadromous (or "up running") fish; born in freshwater streams, they migrate out into the open ocean to mature, then return to those same inland streams to spawn and die. Each species – pink and coho, chinook and chum and sockeye – has its own cyclical rhythm for this oscillation between fresh and salt water. Chinook salmon, for example, spend a year in fresh water and up to five years in the open ocean before making the return journey upriver; pink salmon, after only a few months in fresh water, spend a single winter at sea. Chum salmon take two to five years in the salt of the open ocean, while coho salmon return inland after one and a half or two years. The migratory rhythms of sockeye salmon are the most diverse of the bunch. They can spend three months to three years in fresh water, and from one to four years in the ocean brine. The diversity and differentiation of the sockeye cycles results from the great variety of their freshwater habitats – the different lengths and sizes of their rivers, streams, and lakes.

In fact, any wild Pacific salmon will belong not only to one of the five Pacific species (chinook, pink, chum, coho, or sockeye), but also to one of hundreds of subgroups, a race or "run" unique to its natal river or tributary stream. Each race is marked by particular qualities coevolved with that stream, corporeal traits exquisitely tuned to the characteristics of that watershed, to – for example – the melting glaciers on the mountains above, and the steep or gradual grade of a stream's flow, to the coarse or fine grain of its sediments and the specific chemistry of its waters (itself informed by the geology through which those waters percolate, by the gnarled roots that jut into those waters and the specific leaves that fall upon their surface), to the patterns of shade afforded by the surrounding forest, and the various predators (river otters, eagles, bear) of the place. Each race of salmon is thus the rhythmic pulse of a particular place, each individual an ocean-going avatar of a specific stream or inland lake.

Not so long ago, almost every stream, river, and rill emptying into the North Pacific – from southern Japan to northern Siberia, around the huge Alaskan land mass and the Pacific coast of Canada all the way to southern California – supported one or more salmon runs perfectly adapted to the dynamic ecology of that watershed. The migratory ways of the salmon were likely born in response to the swelling and subsiding ice ages that have dominated their range after these fish evolved to their present form in the cold fresh water of the northern latitudes some two million years ago. As those inland waters were subsumed beneath the immense, spreading ice sheets, the fish were driven out into the ocean and forced to adapt, yet somehow they never lost their ancestral tie to the fresh mountain streams. Whenever the ice sheets retreated, salmon would colonize the rivers and tributaries formed by the glacial runoff, slowly establishing new spawning beds in valleys scoured out and scraped by the ice. They brought with them the rich nutrients of the open ocean, and as bear, eagle, and otter feasted upon their spawning or spawned-out bodies, this abundant ocean nourishment was distributed more widely, enriching the soils and enabling the sparse pioneer woodlands to fill out into dense forests. These forests, in turn, shaded the inland streams, their detritus providing shelter and food for the aquatic insects and small fish upon which the salmon themselves fed. The thickening woods offered habitat for innumerable other animals, for raven and coyote, for owl and deer and raccoon. The association between the salmon and woodlands is an ancient reciprocity renewed time and again: as ice retreated, the fish and the forests recolonized the post-glacial terrain together.

After the most recent ice age, nomadic bands of humans also made their way up into the coastal forests and the river valleys, drawing sustenance from the seasonal storms of large fish that would periodically undulate up the rivers, surging past fallen tree trunks and leaping up waterfalls. Many centuries later, while developing ways of preserving the caught salmon (drying them in the wind or smoking them on wooden racks), some Indigenous peoples settled in permanent villages along those rivers. Through trial and calamitous error – at times overfishing the runs and having to endure the consequent seasons of famine – these cultures gradually learned how best to harvest the collective gift without interrupting its cyclical replenishment. Central to all such cultural constraints, along the Pacific Rim, was a recognition of the salmon as a powerful emissary from hidden or unseen dimensions – a form of energetic intelligence that came towards humankind from the sacred heart of the mysterious. Indigenous cultures from every part of the North Pacific revered the salmon as an uncommonly holy power, ritualizing their respect in ceremonies that honoured the first salmon caught in the spring. No other salmon could be taken during these rites, wherein the first salmon was treated as an esteemed guest before being carefully prepared and eaten.

Along the Klamath River in what is now California, the primary "first salmon ceremony" was conducted in a Yurok village close by the mouth of the river. After that event, strong runners were sent upriver to alert the Hupa that the proper rites had been accomplished, and that the spring salmon were on their way. Upon catching their own first fish, the Hupa undertook ten days of ceremony and prayer before allowing regular salmon fishing to commence. The Karuk people, many kilometres further upstream, moved away from the river and into the hills while the first salmon was taken and ritually eaten by their spiritual leaders.

One effect of such ceremonies, and of the restrictions on fishing during their enactment, was that significant numbers of early salmon were enabled to pass freely upriver to their spawning grounds, ensuring the continued replenishment of the run. The ritualized honouring of the first fish also ensured that the salmon, however abundant in the coming season, could not be taken for granted – that its flesh remained a sacrament for the people.

Across the Pacific, among the Ainu people of northern Japan, whenever any family caught the season's first salmon from the river, the fish was passed through a special window into the house before being placed in front of the hearth fire. There the family would address the spirit of the salmon directly, honouring it ceremonially with spoken words and ritual gestures. The household fire, for the Ainu, was itself a goddess who could see all that unfolded around her; she would report back to the other gods that the salmon had been treated with proper respect.

The Ainu held ceremonies, too, to bid goodbye to the salmon when, having left their flesh bodies behind as food, they paddled their spirit boats back to their homes far to the east. Like other Indigenous peoples of the North Pacific, the Ainu assumed that the salmon, when they were not crowding upstream to visit the people, removed their salmon garb and lived in human form beyond the ocean horizon. Such a mythic view bound their sensory imagination to the ways of this wild creature, engendering an almost familial regard for its well-being. The view was so widespread that when, in the nineteenth century, several Skagit Indians from the American Northwest accompanied a white expedition to the East Coast and saw the abundance of pale, pink-skinned people living there, they reported back to their tribe that *they had been to salmon country and had seen the salmon walking around as human beings* (Collins 1952).

The Hidden Alliance

More detached and technological approaches to tracking salmon have yielded other ways of describing their whereabouts once they depart the inland waters. Upon leaving their rivers, the salmon seem to spend the largest part of their lives swimming in great circles throughout the North Pacific. Their journeys carry them to the remotest regions of the sea, feeding and growing strong on the ocean's abundance – on herring and smelt and other small fish – travelling distances that boggle the human mind. After several years of dispersing to all points on the horizon, following their food whence it leads them, the members of a single run unerringly return to the mouth of their natal stream – all converging there, somehow, at precisely the same time. How they pull off this feat remains an enigma for present-day science. Once the salmon come close to their home stream, it is probable that they rely on their astonishingly keen sense of smell to distinguish between the subtly different waters of neighbouring tributaries. But how the fish navigate across thousands of kilometres of ever-shifting and largely featureless ocean to make their way back to the very same coastal point from whence they set out years earlier remains an elemental mystery to us, confounding our primate senses and our terrestrial, pedestrian logic.

Like spring monarchs fluttering north towards specific clumps of milkweed that only their great-grandchildren will reach, like sandhill cranes vibrating the sky with their bugling as they drop towards a tiny patch of peat bog in the broad tundra, the migrating salmon appear to avail themselves of somatic skills far beyond our bodily ken. The only way contemporary science seems able to fathom their uncanny navigational powers is by likening the abilities of these animals to technologies of our own human invention. We are told, over and over again, that these migratory creatures make use of internal maps and inner compasses, of innate calendars and inborn clocks. Clocks, compasses, and calendars, however, are by definition *external* contrivances, ingeniously built tools that we deploy at will. Metaphorically attributing such instrumentation to other animals has confounding implications, suggesting a curious doubleness in the other creature – a separated sentience or self that regularly steps back, within its body or brain, to consult the map or the calendar.

It seems unlikely, however, that organisms interact with an internal representation of the land in any manner resembling our own engagement with maps. Cranes and butterflies would have little use for a separated *re-presentation* of the Earth's surface, for they have never torn themselves out of the encompassing *presence* of the wide Earth. Our reliance upon such instrumental metaphors seems to stem from our civilized assumption of a neat distinction between living organisms and the non-living terrain that they inhabit, an unambiguous divide between animate life and the ostensibly inanimate planet on which life happens to locate itself. As long as the material ground is considered inert – as long as the elemental atmosphere or ocean is viewed as a passive substrate – then the long-range migrations of certain animals can only be a conundrum, a puzzle we'll strive to solve by continually compounding various internal mechanisms that might somehow, in combination, grant a particular creature the power to grapple its way across the world.

Instead of hypothesizing more metaphorical gadgets, adding further accessories to a crane's or a salmon's interior array of tools, what if we were to allow that the animal's migratory skill arises from a felt rapport between its body and the breathing Earth? That a crane's 3,000-kilometre journey across the span of a continent is propelled by the felt unison between its flexing muscles and the sensitive Flesh of this planet (this huge curved expanse, roiling with air currents and rippling with electromagnetic pulses), and so is enacted as much by Earth's vitality as by the bird that flies within it?

Such a conception need not contradict any of the accepted evidence gathered from a century's research into the enigmas of animal migration; it simply offers a new way of interpreting and integrating those various evidences. By focusing our questions so intently on the organism, as if it carries all the secrets of this magic hidden within itself, we easily lose sight of the obvious collaboration that's at play. By adding new gadgets to an animal's neurological and genetic endowment, we tacitly induce ourselves to focus upon relationships interior to the organism (how, for example, does the animal bring its biological clock and its internal map to bear on its compass readings?), deflecting our curiosity and attention from the more mysterious relationship that calls such interactions into being.

What is this dynamic alliance between an animal and the animate orb that gives it breath? What seasonal tensions and relaxations in the atmosphere, what subtle torsions in the geosphere help to draw half a million cranes so precisely across the continent? What rolling sequence or succession of blossomings helps summon these millions of butterflies across the belly of the land? What alterations in the olfactory medium, what bursts of solar exuberance through the magnetosphere, what attractions and repulsions? For surely, really and truly, these migratory creatures are not taking readings from technical instruments or mathematically calculating angles; they are riding waves of sensation, responding attentively to allurements and gestures in the topological manifold, reverberating subtle expressions that reach them from afar. These beings are dancing not with themselves but with the animate rondure of the Earth, their wider Flesh.

Consider the deep somatic attunement by which a salmon feels its way between faint electromagnetic anomalies, riding a particular angle of sun as it filters down through the rippled surface, gliding with certain currents and plunging up against others, dreaming its way through gradients of scent and taste towards a particular bend of gravel and streamside shadow. Whatever specialized sensitivities and internal organs are brought to bear, those very organs have coevolved with textured patterns and pulses actively propagating through the elemental medium; indeed, those sensitivities have often been provoked by large-scale repetitive or rhythmic happenings proper to that part of the biosphere – by pulsed coalescences and cyclic dispersals – and so can hardly be fathomed without reference to these patterned gestures within the body of the planet.

Perhaps it would be useful, now and then, to consider the large, collective migrations of various creatures as active expressions of the Earth itself. To consider them as slow gestures of a living geology, improvisational experiments that gradually stabilized into habits now necessary to the ongoing metabolism of the sphere. For truly: are not these cyclical pilgrimages – these huge, creaturely hegiras – also pulsations within the broad body of Earth? Are they not ways that divergent places or ecosystems communicate with one another, trading vital qualities essential to their continued flourishing?

Think again of the salmon, this gift born of the rocky gravels and melting glaciers, nurtured by colossal cedars and by tumbled trunks decked with ferns, fungi, and moss, an aquatic, muscled energy strengthening itself in the mossed and forested mountains until it's ready to be released into the broad ocean. Pouring seaward, it adds itself to that voluminous cauldron of currents spiralling in huge gyres, shaded by algal blooms and charged by faint glissandos of whalesong ... Until, grown large with the sea's abundance, this ocean-infused life flows back up the rivers and tributaries and spreads out into the wooded valleys, gifting the hollows and the needled highlands with new minerals and nutrients, feeding bears and osprey and eagles, ensuring that the glinting gift will be reborn afresh from the lump of luminous eggs stashed beneath a layer of pebbles.

This circulation, this systole and diastole, is one of the surest signs that this Earth is alive – a rhythmic pulse of silvery, glacier-fed brilliance pouring through various arteries into the wide body of the ocean, circulating and growing there, only to return by various veins to the beating heart of the forest, gravid with new life.

Or ... Perhaps it's better to think of this seasonal reciprocity as a kind of *breathing*, as an exhalation of millions of young salmon smolts down from the tree-thick mountains and meadows and then out into the roiling cosmos of currents and tidal flows, to mingle with zooplankton and seals and squids, and then the great in-breath, the drawing in of living nourishment from the sea into river mouths and estuaries, inhaling the salmon up the rivers into streams, and from there into the branching becks, rills, and runnels that filter into the green forests, the living lungs of this biosphere. Or is it the broad-bellied ocean that is breathing, sucking these finned nutrients down from the shaded slopes, luring them over rocks and through rapids and hydroelectric dam spillways – drawing them past bustling cities and factories, through intersecting gradients of toxic effluents that sting their mouths and strafe their exposed membranes – on out into the heaving whirl of the sea's innards, circulating this glimmering nourishment within itself before exhaling it back, a long, sighing breath, up into the wooded valleys?

However else we may view them, these deterritorializations and reterritorializations – these large migrations of various species – are a primary way that the biosphere cleanses and flexes its various organs, replenishing itself, each region drawing insight from the others, concentrating and transforming such qualities before releasing them abroad, divergent places trading perspectives along with nutrients and nucleotides, the whole half-shadowed sphere steadily experimenting, improvising, slowly altering its display to the blazing fire watching from afar, as the reflective moon rolls on 'round.

The Outer Bound of Oblivion

Might not humans, too, recognize ourselves as expressions of the animate Earth? Are we not fluctuations within the broad body of the biosphere? Can we not feel many of our collective behaviours as dynamic gestures proper to the planet itself? Perhaps. Yet this shift of perspective is, in our case, much more difficult to accomplish. After all, much of the hustle and bustle of contemporary civilization seems entirely contrary to the continued flourishing of the biosphere. This roiling tumult of humans hurtling every which way across the land, and the accompanying frenzy of cutting, clearing, damming, paving, and building, is tough to reconcile with the wider, more-than-human commonwealth, especially when so many other species choke on the byproducts of all this human "development" – their flocks dwindling, their movements stymied, their communities crowded out by our unending profligacy. Our own weedy, invasive species had been slowly growing for many long millennia, yet we've become a full-blown eruption since we began burning huge amounts of fossil fuel. In little more than a century, we've expended vast reserves of stored sunlight that were laid down and concentrated, deep in the Earth's crust, over hundreds of millions of years. The extravagant swelling of our numbers, and the crazed speed at which we two-leggeds are conducting our lives – the headlong chaos of our manifold trajectories, along with our rapidly dwindling attention spans – is tightly bound to this conflagration of stored sunlight.

Modern humankind would seem to be on a kind of drunken binge, unable to moderate our craving for this instant and overwhelming power, careening this way and that, smashing the fixtures and trashing the house. More broadly, we could say that the steady injection of cheap oil into our veins seems to have induced a prolonged spawn within our species, an unending expansion of our population without regard for what surrounds us. I can't help but think of the spawning salmon so oblivious to my presence among them as they bumped into my naked knees and plunged past, nosing aside even their dead or dying peers as they single-mindedly surged upstream. Unlike the cyclical replenishment of the salmon, however, whose seasonal pulse nourishes countless other animals and plants, our species seems locked in a spawn that never ends. Our inability to notice the other creatures among us, our readiness to shove aside the many of our own kind who are ailing or dying as a result of our recklessness, seems never to abate.

Are we the biosphere's way of exhausting itself, of bringing on a morbid fever, vomiting precious stores of vitality into the thickening atmosphere?

Like the olfactory sensitivity of salmon, like the keen eyes and ears of wolves, our human senses have coevolved with the enfolding Earth, and hence are dynamically tuned to variant aspects of the terrain vital to our continued well-being. Sensory perception binds each individual nervous system into the encompassing ecosystem. (For each creature has its own *chiasm*, its own reciprocal interchange with the wider flesh of the world.) Yet as more and more tools have inserted themselves between our bodies and the living land, the human nervous system has been progressively decoupled from the wider interplay of biospheric forces. As the spontaneous reciprocity between our senses and the earthly sensuous has been short-circuited by our increasing engagement with our own technologies – as we've entered into a closed loop with our own creations – humankind has slipped into a kind of freefall or runaway state, unconstrained by our actual surroundings.

Oil has a concentrated charisma very difficult to resist. Cheap oil powers the manufacture of our digital artifacts; cheap electricity from fossil fuel enables the functioning of those many gadgets, powering the immense webwork of electronic exchange. With our animal senses transfixed by our own human-generated media – our ears gripped by the words or rhythms churning from our headphones, our visual focus captured by the images steadily dancing across our handheld, desktop, and wall-mounted screens – there's precious little chance for the wider, more-than-human world to break through our collective trance. And yet it *is* breaking through, every day, and all around us. However completely we've cocooned ourselves within a virtual world of our own making, our animal body still takes its final directives from the larger biosphere – it still needs to breathe and to drink clean water. It remains susceptible to the tremor of earthly wonder, to creaturely empathy, to illness. As planetary temperatures rise ever higher, as local weather patterns go haywire, the unseen medium of air that conjoins our awareness with the sentience of cedars, sandhill cranes, and spiders becomes more insistently evident, more palpable, muscling its way back into consciousness. As water sources dry up and disappear, as transportation lines are interrupted by climatic upheavals, as crops fail and infections spread, as economies falter and rolling power outages become more commonplace, we look up from our flickering screens, trying to fathom what's unfolding. And so find ourselves immersed, once again, in the terrifying beauty of a world that exceeds all our knowing.

Is this biosphere our realer Body, the broad metabolism in which our smaller, more transient bodies are entangled? Is this Earth not the very Flesh of which my own flesh is woven; am I not a steady interweaving of this more intimate flesh, here, with that wider Flesh all around? Does not our language, which Merleau-Ponty teaches us to discern as a corporeal field of stuttering exuberance and desire, fall under the influence of gravity? Does not our discourse, this sonorous sounding or singing to one another – overlaid these days by our feverish tweets and twitterings, and underlain by the thrumming hum of our power plants, but still – does not our collective and sometimes cacophonous speaking have the topology of a sphere? Is it not this sphere that is singing through us? Is not our real intelligence the palpable presence of these leafing forests, of those mountains conjuring clouds out of the fathomless blue, of these tides laced with crude and strewn with bright-coloured plastic? Have we yet reached the outer bound of our oblivion?

NOTE

A very different version of this chapter was published in both textual and audio form in the inaugural issue of *Emergence Magazine* (Spring 2018), an online quarterly, and can be found here: https://emergencemagazine.org/story/creaturely-migrations-breathing-planet/

REFERENCE

Collins, June. 1952. "The Mythological Basis for Attitudes Towards Animals among Salish-Speaking Indians." *Journal of American Folklore* 65: 353–9. https://doi.org/10.2307/536039.

4 When Salmon Are Deemed Superfluous: Reflecting on a Struggle of Stories

MARTIN LEE MUELLER

Norway, 2010

In August 2010, Professor Rögnvaldur Hannesson of the Norwegian School for Economics and Business Administration published an opinion piece on the future of Norwegian salmon in the business newspaper *Dagens Næringsliv*. His article "What Shall We Do with Wild Salmon?" asked whether the time had come to sacrifice all of Norway's wild salmon in favour of their domesticated cousins. The newspaper described Hannesson as one of the country's leading experts on fishery economics, and so his contention held a certain authority. Here is what Hannesson wrote:

> We should perhaps ask ourselves what we want wild salmon for. If wild salmon get in the way of the fish farming industry, then I must say we must be ready to sacrifice wild salmon. The industry creates great values and jobs along the entire coast. It is an important business branch, one that is important to keep. We need not feel pity for the upper class that will miss a playroom; surely they'll find some corresponding amusement. (Hannesson 2010; my translation)

The Pacific Northwest, 1910

A century earlier, businessman Thomas Aldwell began constructing the first of two hydroelectric dams that would block the Elwha River on the Olympic Peninsula. Historically, the Elwha was among the few rivers in the contiguous United States to house all of the anadromous salmon and trout species native to the Pacific Northwest. Aldwell envisioned that the dams would yield "peace, power, and civilization" (1950, 80). The two dams were widely recognized to be a significant step towards modernizing the American West. They would convert the Elwha "from its waste and loss into a magnificent source of energy and strength," as the *Sequim Press* put it at the time (14). Journalist William Ware from the *Seattle Post-Intelligencer* had written the following as early as 1 December 1901, under the headline "Scenic Wonders of the Picturesque Elwha River":

> The grandest river of its size in Washington, beautiful, clear as crystal, rushing down from the snowcapped peaks of the majestic Olympics, through gorges, over cataracts, through and among immense boulders, cool and pure and powerful, containing the energy of thousands upon thousands of horsepower, it carries along upon its crest the mighty monarch of the forest as easily as it does the tiny Indian canoe. The Elwha, sublime in its majestic and awe-inspiring scenery, is destined to become a mighty power for good in the hands of ingenious humanity, for the present and future generations ... There is no river in Washington with the grand future before it, with its immense possibilities, with the undeveloped energy or with the natural facilities that this river has. There are many places along its course where its energy could be transformed into power for the use of the manufacturer, for lighting, for the tramcar, for the streetcar. (Mapes 2013, 53; emphasis added)

In the hands of ingenious humanity, this mighty power for good was constructed without fish ladders that would permit the salmon to climb past the dam site. And while, a year after construction had begun, the *Seattle Times* announced proudly that "[the] Elwha is now under control" (Mapes 2013, 56), the fish – once renowned as the mightiest salmon in the Olympics – began to decline. By the early twenty-first century, their numbers had dwindled from a historical 400,000 annual returnees to less than 1 per cent of that number.

Though a century apart, these two cases have some striking similarities. Here as there, salmon are deemed superfluous and in the way of industry. Here as there is the notion that sacrificing wild salmon is a perfectly rational choice, an inconvenience at best. Here as there, the more-than-human world of rivers, oceans, and their many plant and animal inhabitants is thought of as a purveyor of raw materials. This kind of rationale is still dominant today, and has been so since Descartes first codified such thinking philosophically in the seventeenth century. In metaphysical terms, rivers and oceans had become *res extensa*, a mechanistic collection of objects rather than a communion of living subjects, as Thomas Berry put it (Waldau and Patton 2006). This kind of metaphysics helps redefine exploitation into something inherently good and necessary.

Norway, Today

Prime Minister Erna Solberg's administration recently published a report on the future of wild and farmed salmon along Norway's vast and beautiful coastline. Titled *Foreseeable & Environmentally Sustainable Growth in Norwegian Salmon and Trout Farming*, the report defines an ambitious goal for the coming generation: Norway's salmon production – already world-leading, with over one million tonnes of salmon flesh shipped to foreign markets annually, which corresponds to about 500 million individual salmon slaughtered – is set to grow five-fold by mid-century. A careful reading of the report shows that mechanistic, hierarchical thinking by no means belongs to the past, nor does it merely reflect the isolated opinion of one eccentric, though undoubtedly influential, researcher. The report shows that in the case of Norway, such thinking remains foundational for official government policy concerning the fate of salmon.

While the country's number one export industry, oil and gas extraction, is increasingly coming under pressure by an ecologically literate and alert public, much political ambition and economic muscle is now being directed to the country's second most successful export industry, aquaculture. Like its oil twin, the aquaculture industry is commonly spoken of as a "fairy tale" or "adventure" (*Lakseeventyr*), a sweeping success story that gives Norway a uniquely competitive edge in the globalized market economy. Solberg said on national radio that "salmon is for Norway what IKEA and H&M are for Sweden – our prime international brand. It is the number one product people abroad associate with our nation" (NRK P2 Radio; my translation). Lisbeth Berg-Hansen, Minister of Fisheries during the previous Stoltenberg administration, boldly declared the aquaculture industry to be part of "the Norwegian identity" (Kristoffersen 2011). Former prime minister Jens Stoltenberg himself repeatedly allowed journalists to report on him while he was advertising farmed salmon during his diplomatic travels abroad, and members of the royal family have released press images from international travels showing them wearing aprons and holding stainless steel knives, with farmed salmon fillets laid out in front of them, ready to be sliced.

In a foreword to the report, Professor Atle Guttormsen of the Norwegian University of Life Sciences writes: "The Norwegian salmon industry is today in a situation where … it meets a global market which screams for more salmon" (Meld. St. 16 [2014–2015], 4). Inflating the industry five-fold to approximately 2.5 billion salmon slaughtered annually is portrayed as a response to a global market which calls rather desperately for more animal protein. The decline of oil and gas extraction may ultimately be inevitable, but here is an opportunity not yet fully grasped. Or so it is claimed. Such perceived need then sets the tone; it is harnessed to create a strong motivation to expand the industry.

And yet the report does caution that future growth must proceed with restraint and care. It establishes two guiding principles that will henceforth serve as baselines for future expansion. First, the report defines what we might think of as a *normative baseline*:

> Norway's coastal and open-sea territory, as well as its coastal streams, are among the most important habitats for wild salmon ... We house about 1/3 of the total population of Atlantic salmon, spread across about 400 individual populations. Norway therefore has *an exceptional international responsibility* for the wild salmon. (Meld. St. 16 [2014–2015], 36; my translation)

Second, the report defines what we can think of as an *ecological baseline*:

> If the salmon industry is to be guided by a predictable growth-policy, we must ... define what environmental impacts the public should accept. The government thinks that *ecological sustainability must be used as the most important guiding principle* for any further growth inside the industry ... [Therefore] the government suggests a moderate risk profile ... (Meld. St. 16 [2014–2015], 8/83; my translation)

The twin baselines read as if they establish a firm and definitive grounding, one that will help navigate competently and alertly through politically charged territory: Norway recognizes its exceptional international responsibility for wild salmon, and it declares that ecological sustainability will be the single most important guiding principle ahead. So far, so good. But there is also that apparently inconspicuous afterthought: *The government suggests a moderate risk profile.* The sentence comes in the wake of that dramatic opening scenario of a desperate global market, and also of valiant claims to moral integrity and ecological farsightedness. Somehow, then, the sentence is spoken rather easily; the call to moderation seems grounded, temperate, sensible. Or does it? What exactly does the Norwegian government mean by a "moderate risk profile"? Here is how the report defines the phrase: "It is likely that 10–30 percent of the [wild salmon] population dies [every year] because of infection by sea lice" (Meld. St. 16 [2014–2015], 60; my translation).

If it is not immediately clear how one may reconcile an "exceptional international responsibility for wild salmon" and "ecological sustainability" with the above definition of "moderation," then that may be because the rhetoric works directly to diffuse a clearer and more critical understanding. Both the government's straightforward recognition of Norway's moral responsibility and the invocation of sustainability as a beacon for further action gradually neutralize the reader's critical alertness. Surely, when the government so clearly recognizes its moral obligation to the international community and to sustainability, one can accept a "moderate risk profile" for action? Well, not necessarily. As soon as we strip the 90-page report of its obscuring rhetoric, its essence comes forth more clearly. And that essence no longer reads so reassuringly: in the interest of further growth across the next human generation, the public should accept the sacrifice of *one third* of all of Norway's wild salmon, year after year, for forty years to come.

It is difficult not to have a historical déjà vu. Did this logic not already make itself felt on the Elwha River a full century ago? There, too, wild salmon were drawn into the logic of a profit-driven commodity market and became, ontologically speaking, externalities: calculated, inevitable losses. And yet something is new here, too: gone is the conspicuous arrogance, such openly anthropocentric tropes as "taming" or "controlling" rivers. Gone also is the boldness and directness with which the fisheries professor suggested letting Norway's salmon become extinct. Here is an arrogance that is rather more subtle and, in a sense, far more subversive: the salmon's calculated death, or planned obsolescence, has become discreet and ordinary. The annual death of one third of all wild salmon has been dressed up in inherently good and, indeed, beautiful language, to the point where the boundaries between right and wrong seem to blur altogether – even when we look directly at what is ultimately a death sentence for hundreds of thousands of wild creatures.

The report invokes the authority of some of the very concepts developed in recent decades to *resist* the ongoing spread of the metaphysics of human dominance, and of the market logic, into the realm of living beings. Here, an eminent *moral* recognition, distilled laboriously by the philosophical community: Norway, by virtue of being the birthplace of so many Atlantic salmon, has an international obligation to protect these fish. Therein lies the no-less-eminent *ecological* recognition that sustainability – the ability to live in place, indefinitely – has come to be understood as a baseline for any human endeavour, including the realm of economics, law, or politics. It is because the concepts are being eroded

so candidly, right there in the open, that their erosion becomes potentially so effective, and the calculated death of wild salmon becomes so
difficult to resist, both conceptually and on the ground. By virtue of giving the appearance of full transparency, and of democratic legitimacy,
the Norwegian government succeeds in making the deeply troublesome
core of the matter virtually invisible: the officially sanctioned mass dying
becomes *fully* transparent indeed.

The Elwha River, Today

The recent dismantling of both Elwha River dams has made headlines
as the largest dam removal so far in history, anywhere. Dams once built
with the explicit intention to expand industry, and invested with such
ambitions as to advance matters of peace, power, and civilization, have
been dismantled, giving the salmon in the river the opportunity to
rebound from near extinction. The Elwha's breakthrough moment
came in 1992. Indigenous tribes and environmental groups had pe-
titioned for restoration of the river and its salmon runs. That year,
their petitions were heard. President George H. Bush signed into law
the government's decision to buy the dams, setting in motion a series
of studies concerning dam removal. From the moment Bush signed
the legislation to the moment when the dams were blasted out of
the river, more than two decades would pass. But in August 2014, the
last remnants of the dams were gone. And just two weeks later, after
a 100-year-exile, salmon were already moving upstream again beyond
the former dam site.

When the two dams were built a century ago, a law mandated that
every new dam site must be given a fish ladder, so that homecoming fish
could move up around the dam and onward to their spawning grounds.
The engineers on the Elwha decided the ladder was too expensive to
build, and a small hatchery that was built instead was soon closed down
again. Criminal charges were never laid, though a law had clearly been
breached. And so for a hundred years, the dams stood as a physical articulation of that Cartesian metaphysics that deemed the Elwha River
a mere purveyor of raw energy, and its migrating fish superfluous – a
somewhat unfortunate but inevitable side effect of progress.

But something else was underway in the Elwha watershed, something
rather elusive, tenuous. For a full century, even as the insurmountable
Elwha Dam rose from bank to bank, *some* salmon still kept on jumping
against that concrete wall, testing it for opportunities to leap across it, year
after year. They simply didn't give up trying. They rather unmistakably
articulated a staunch determination to go on returning their charged

bodies to the upriver spawning beds that their elders had once taken for granted.

Recent events on the Elwha River may be extraordinary in terms of both scale and audacity, but they are not isolated. Rather, they spearhead a development that has recently seized the Pacific West Coast as a whole, from California in the south to Alaska in the north. Across all of the Pacific salmon's historical spawning range, there is now evidence that a truly more-than-human alliance is struggling to emerge, an alliance between Pacific salmon and the human inhabitants inside the many watersheds of the region. Some even think of the region as Salmon Nation, suggesting a collective, multi-ethnic sense of belonging imprinted on the region by the different species of Pacific salmon (cf. Wolf and Zuckerman 2003).

This more-than-human alliance is now attempting to create more respectful, ecologically sound, and economically diverse human–salmon relations. Likewise and in equal measure, the civil society of that region is dedicated to trying to bring the various articulations of human culture into a certain resonance, or reciprocity, with the ways of salmon. Countless local projects have sprung up in recent years, involving rural communities and cities, scientists and schoolchildren, businesspeople and poets, Indigenous leaders and public servants (cf. House 1999; Duncan 2001; Wolf and Zuckerman 2003; Mapes 2013). This more-than-human alliance between humans and salmon boasts successes in economic innovation towards less extractive, less arrogant, and more reciprocal ways of receiving the gift of salmon (cf. Wolf and Zuckerman 2003). It has encouraged a number of recent scientific insights into the great importance of salmon as a keystone species for the larger bioregion, including remarkable achievements in ecology and biology, but also including fascinating descriptions from ethnographers, anthropologists, and wisdom keepers (both Indigenous and other) about the subtle ways in which salmon enrich not only rivers or forests, but also the collective human imagination (cf. Nelson 1991; Jensen 2004; Johnsen 2009; Crane 2011; Colombi and Brooks 2012; Deloria 2012; Mueller 2017).

In all this, the dismantling of the two Elwha dams marks a concrete political act of *restoring* a landscape, but it also marks an important symbolic gesture, an act of what phenomenologist David Abram thinks of as *restorying* (cf. Abram 2010): the dismantling of the dams has initiated a re-examination of the various peoples' complicated relationship with the larger living community, and salmon are increasingly recognized as being the keystone to this inter-ethnic work of restorying. They are being recognized as creatures deeply entangled not only with the ecology but also with the *mind* of the Pacific Rim. Salmon are beings of flesh, blood, intention, sentience, and intelligence, but they are also symbolic creatures, totemic beings who nourish the human imagination with insights, metaphors, wonder.

The Elwha case symbolizes defiance, determination, and also love for the strange and exuberant otherness of salmon. And it symbolizes a striving to recreate a more complex, reciprocal, integrated, and beautiful relationship between humans and the more-than-human world. There, as elsewhere across the Pacific Northwest, people are asking: What are the needs of the salmon in these streams? What are the needs of those rivers, and the many other creatures that also depend on salmon flesh for their lives? Further: How can the multi-ethnic groups of humans inside the many watersheds live in such a way that they once again become accomplices of the land, rather than disturbances? Those are questions one now encounters again and again across Salmon Nation, and the chorus of defiant and devoted voices who challenge the anthropocentric story is still swelling to a crescendo (cf. Woelffe-Erskine, Cole, and Danger 2007).

A Generational Struggle

But, sure enough, the region, like Norway, struggles with a powerful and profit-driven salmon industry and, in metaphysical terms, with the difficulty of containing and unweaving a rampant story of human exceptionalism. Still, that story clings stubbornly to the outdated notion that humans ought to be at the centre of the living web and the very apex of all beings. The same year that Norway's government published its report on how to increase its feedlot industry five-fold, the US Food and Drug Administration (FDA) certified genetically modified salmon for trade in the US (FDA 2015). The FDA's approval of the so-called AquAdvantage Salmon™ earned salmon the dubious distinction of being the first genetically modified animal ever to be approved for sale. Likewise, farmed *Atlantic* salmon are now British Columbia's largest "agricultural export" (Cermaq 2014), produced in feedlots owned largely by Norwegian companies – planting this keystone of corporate Norwegian identity into fjords and inlets far from home. Inevitably, conflicts between British Columbia's feedlot industry and local residents are on the rise (Davis 2016; Clayoquot Action 2015; Morton 2019).

And yet: Seen from Norway, which the author of this chapter calls home, the Pacific Northwest as a whole appears as a throbbing real-life laboratory where matters of philosophical concern and matters of public policy are equally being enlisted to pierce, and ultimately disable, the archaic story of human domination, and with some astonishing successes. This is worth pointing out, for when it comes to salmon, in Norway there is no civil society dedicated to the survival of salmon that could compare in terms of innovation, momentum, and success. In some sense, Norway's political and economic elites hold the country captive in a time loop, say, in the Elwha River's dammed year, 1910. In some sense, those of us here who labour to speak and act on behalf of a more-than-human sensitivity have a century to catch up on when it comes to mobilizing effectively against the story of human supremacy.

This is in no way intended to underestimate the struggles and difficulties that lie ahead for the Pacific Rim. Both there and here in Norway, the story of human domination has suffused the modern lifeworld in ways that are thorny, resilient, and ubiquitous, reaching into the legal, political, economic, and scientific imagination, propagating itself through technology as well as through social institutions, resounding even in grammar or particular speech habits, and subtly shaping even the ways in which we humans inhabit space and time. Whether we live on the Atlantic Rim or the Pacific Rim, the work of resisting that story, and of reweaving more life-affirming and reciprocal narrative strands with the salmon and the larger living community, must ultimately address all of these complexities.

Evidently, this is a generational struggle, limited neither by national boundaries, nor ethnic allegiances, nor cultural heritage. First and foremost, it concerns us all as fellow human beings. For the time being, the dominant narrative is still able to legitimize its claims to truth and authority. By virtue of institutional power, and of metaphysical habit, it can still circumscribe what is "normal" and what is "deviant," which ways of knowing are tolerable and which are inappropriate, who is invited into the circle of moral concern and who is kept outside. And yet, and again, the chorus of defiance is gaining momentum, and it is forging alliances across perceived ethnic boundaries as well as across species boundaries.

Cursory though these notes are, they do seem to suggest that questions concerning the charged human–salmon relationship are inseparable from questions of power. And questions of power are inseparable from struggles over narratives. Valuable concepts such as international responsibility for the well-being of wild creatures or sustainability have the potential to truly challenge existing power arrangements and to destabilize the dominant narrative of human exceptionalism. But the example of the Norwegian government's report shows how effectively power elites can absorb blows to their authority, and how easily good tools can be co-opted and rendered ineffective, to the point where once-useful tools are turned against what they were created to critique. To sacrifice hundreds of thousands of wild salmon in the interests of sustainability seems to be the ultimate travesty of the very essence of sustainability.

The fact that a democratically elected government is nevertheless able to proceed with it, in bright daylight and without any effective resistance, indicates some of the difficulties that lie ahead.

Questioning the Legitimacy of the Anthropocentric Story

What one may wonder is this: Is it possible for us not only to acknowledge but actually to disarm and neutralize such dynamics? My hunch is that to answer this question, we can turn to another one: What gives the foundational story of human supremacy legitimacy in the first place? Can we not reverse the burden of proof and ask: What ground does the anthropocentric story have for its claims to authority and power? In some sense, then, our reversal of the burden of proof turns this conversation into one about strategy, and for a reason: when we frame the smooth and often shameless co-opting we have seen as a problem concerning the very foundations for legitimacy and authority, we no longer simply react to problems defined within a certain frame of speaking and thinking. We sooner authorize ourselves to redraw the very boundaries of that frame.

My approach to those questions is, as will become clear, a phenomenological one. I do not think that such an approach alone can fathom these questions in their necessary complexities, let alone do so in the span of a few pages. But I do think that such a phenomenological approach can illuminate some of the possibilities available to us as we work to reframe the question of legitimacy. With this in mind, I wonder: What happens when we no longer take the dominant story's claim to legitimacy for granted?

To begin with, it seems prudent simply to observe that the story's legitimacy is largely self-proclaimed. Though ubiquitous, it is neither natural nor normal, nor is it invulnerable to critique. Next, it also seems prudent to observe that there is no such thing as neutral or objective knowledge, no view from nowhere. The salmon feedlot industry repeatedly claims legitimacy by resorting to "scientific knowledge," the implication being that scientific knowledge is beyond and before any ideological struggles, a neutral epistemological ground. This assertion, of course, is itself ideological. It, too, is part of a narrative whose roots reach back into that early modern separation of the thinking mind from embodied ways of knowing, from feelings, intuition, and the sensuous sensitivities of our breathing bodies. It is a narrative that habitually values the quantitative over the qualitative, abstraction over immersion, centralized ways of knowing over place-specific and embedded ways of knowing. If this narrative feels legitimate, if it in some ways seems normal or commonsensical, then it is not because it is so in any ahistorical, objective, a priori sense. Rather, it has created a lifeworld so pervasive as to appear nearly universal, nearly inevitable. The narrative has in some sense become invisible as narrative, like the air we breathe. Fully transparent.

The next observation to hold on to as we question the story's legitimacy is that the thinking mind – though undoubtedly a precious and useful helper – seems strangely limited in its ability to truly challenge the narrative's claim to legitimacy. It seems that there is only so much we can do by more interpretation, more analysis, more thought bending. This observation is in no way intended to disclaim thinking; it is merely to suggest that rationally derived argumentation on its own may somehow not be enough to help us effectively reframe the conversation. What to make of this? Let us consider the following lines by the Oregon-born novelist and fly fisherman David James Duncan. This is from his memoir, *My Story as Told by Water*:

> Reverence for life is the basis of compassion, and of biological health. This is why, much as it may embarrass those of us trained in the agnostic sciences, I believe every life-loving human on Earth carries a far-from-agnostic obligation to remain primitive enough, and reverent enough, to stand up and say to any kind of political power or poll or public: *Trees and mountains are holy. Rain and rivers are holy. Salmon are holy. For this reason alone I will fight with all my might to keep them alive.* (2001, 107; emphasis in original)

Duncan adds that such speaking "is not an argument, not a number, not a polled opinion"; it does not seek to convey thought-derived essences. A reading of his work seems, rather, to suggest that his is a form of speaking that originates through – and draws its authority from – other sources. They include, in his case, a lifelong immersion into the topographies of Pacific salmon and trout streams, wading for an entire childhood, youth, and adult life through rivers sometimes visited by returning fish too numerable to comprehend, and sometimes haunted by absences too painful to overlook. They include also a bodily sensitivity and intellectual vigour sharpened, again across the span of a lifetime, by encounters with grebes, owls, elk, temperate rainforests, volcanoes, salt lakes, cutthroat, steelhead, and, again and again, salmon. Where Duncan speaks with authority and weight, my impression is that he does so not by claiming legitimacy, but rather by bringing the land itself alive through him. His authority is not assumed; it is bequeathed. He speaks not with his voice alone; instead, his writing gathers and makes space for a refracted multiplicity of animal, rock, and water voices. Each such voice then, through his pen, asserts itself in its own significance. At one point Duncan writes:

> We may be elk at heart, forced to live like sheep, or salmon forced to live like carp. The blood still begs direction home. No matter how circumstance

has dealt with us, or how disenfranchised or placeless we've become, we will never be dispossessed of the right to feel at home in *these* little blood-filled bodies, on *this* reeling planet, as often as we can, as deeply as we can, any god-damned God-blessed place we possibly can. (2001, 56; emphasis in original)

What Duncan pits against the self-proclaimed legitimacy of the narrative of human dominance is his work to speak clearly, unflinchingly, and matter-of-factly of the wider and wilder life he finds himself thrown into. His work is the ongoing attempt to invite other Earth voices to pierce the self-inflicted isolation, to assert themselves in our midst, each with its unique blend of fragility and resilience, specificity and universality, splendour and imperfection. We may go so far as to say that the resulting authority – it takes some confidence to declare, "Trees and mountains are holy. Rain and rivers are holy. Salmon are holy." – is more solid and legitimate than any form of self-proclaimed legitimacy could be, because it is ratified not through exploitation but mutual alliances, and it is held in precarious check by numerous more-than-human intentions, sentiences, and lives. Duncan does more than deconstruct the dominant story's self-proclaimed legitimacy through a succession of arguments. When he engages mind and body in his poetic work, when he labours to let human and other-than-human resonate through his pen, when he writes against the tendency to create opposites in the first place, he creates the very atmosphere in which legitimacy, and authority, can begin to reveal themselves. Here is one more invocation from his memoir:

The gifts Nature keeps trying to bequeath us are astounding, if we simply greet those gifts with a world that enables them to be ...
  Salmon are a light darting not just through water, but through the human mind and heart. Salmon help shield us from fear of death by showing us how to follow our course without fear, and how to give ourselves for the sake of things greater than ourselves. Their mass passage, from the sea's free invisible into the river's sacrificial and seen, is not just every American's, but every Earth-born man, woman, and child's birthright. Their bodies remain the needle, their migration the thread, that sews this vast, broken region into a whole. No kilowatt can replace this, no barge can transport it. The Columbia that Industrial Man has given us is dying. The rivers least touched by man thrive. The finned, winged, and four-leggeds watch, waiting to join us, or not, in the world we do or do not create. (2001, 113)

There may in fact be a different and more comprehensive way of knowing that would defy the perceived legitimacy of the narrative of human dominance more effectively, and more directly, than rational thought left on its own.

Forging Different Kinds of Alliances

Again, this is no critique against rational thought as such. It merely suggests that as we ponder questions of narrative, power, and strategy, it makes some sense that we ask again what existing and latent alliances we can identify in this larger generational struggle to preserve life. The notes I have gathered here would seem to suggest that among the most potent alliances that deserve more of our attention are, first, a revitalized and reinforced bond between our embodied minds and our mindful bodies, and second, a strengthened curiosity for the ways in which we – breathing, inquisitive, storytelling two-leggeds – partake of a truly more-than-human world, a life-giving sphere so much vaster, more surprising, and more complex than the received ideology of human dominance would have it. In this approach, I suggest, lie possibilities for forging alliances that could inspire actions not even thinkable from within the frame of the dominant narrative.

Some may find this reading of current events in the Pacific Northwest overly optimistic or even simplistic. I hope to have at least indicated that I am not ignorant of that region's very serious problems. And yet, having staked my professional life here in Norway for a number of years now on the fate of salmon, I dare say that inspiration and encouragement for more decent, beautiful, and reciprocal ways of living alongside these beautiful creatures have consistently (though surely not exclusively) come from the Pacific Rim. The boundaries even of what seems possible are being tested and expanded in that region, through means as diverse as policy making, storytelling, staging acts of civil disobedience, making films, daylighting streams, writing philosophy books, painting murals, hosting tribal canoe journeys, or fuelling up a fleet of government-sanctioned bulldozers to dismantle some of those century-old dams.

And sure enough, for as long as we greet salmon with a world that enables them to thrive, they will keep returning. Being keystone species, they are known to help a veritable crowd of other beings flourish, including grizzlies, cedars, vines, eagles, deer, ducks, owls, spruce, and our own kind. What we witness along the Pacific Rim is that, to some degree, other Earth voices are increasingly finding opportunities to infuse attentive humans with possibilities for new life yet to come and new alliances yet to forge.

Bearing Witness as More-Than-Human Voices Emerge into Our Midst

And so, from time to time, some such Earth voices will resound through the writings of those whose work it is to bear witness.

There is *Seattle Times* writer Lynda Mapes, a long-time chronicler of the events on the Elwha River and the Lower Elwha Klallam community, who describes a visit to the river sometime after the dams have been dismantled. Mark the way in which her voice, not unlike that of Duncan's, appears to be awash in river life. Here, too, we can trace some of that ongoing work, that forging of more-than-human and inter-ethnic alliances, which I invoke in these pages:

> Water tumbled over a branch, driving silver bubbles of fresh oxygen into a plunge pool of cold, clean water ... Here, drama played out on a small scale, as tiny sticks and bark caught in jams of branches that made eddies in the flow. A juvenile fish no longer than the joint of my finger wiggled into the pool beneath the branch, facing upstream as a cluster of alder cones swished by overhead on the water's surface. Threads of algae streamed like hair in the current as the water eased downstream, alive with drifting seeds, detritus, and bugs ... I realized as I watched that the woven pattern of the current, visible in the surface of the water, was mirrored in the stem and vein pattern of the alder leaves, the stem and branch pattern of the sword ferns, and the tiny branched shapes of the moss on the banks. Land and water interwove and interacted at every scale and dimension of the Elwha ... In the return of wholeness to the landscape, tribal members say, is the return of wholeness to the people. Not only for the tribe, but also for the rest of us. (Mapes 2013, 25–7, 10)

Others, too, bear witness to that work. This is the writer Richard Manning, who reflects on a visit to a salmon creek that feeds into Willapa Bay, Washington:

> To those who presume to decide which of the links in the food chain can be sacrificed, there is only one reply: No form of life is dispensable; there is no expendable link ... there is no "other." All of us – salmon, bear, person, tree and stream – are made of the same matter and are in this life together. (Manning 2004)

Then there is Freeman House, late resident of the Northern California Mattole River watershed, who has penned the extraordinary memoir *Totem Salmon.* At one place towards the beginning of his book, he takes the reader right into the river with him, evoking the life of the mind and the body in equal measure:

> King salmon and I are together in the water. The basic bone-felt nature of this encounter never changes, even though I have spent parts of a lifetime seeking the meaning and puzzling over its meaning, trying to find for

myself the right place in it. It is a *large* experience, and it has never failed to contain these elements, at once separate and combined: empty-minded awe; an uneasiness about my own active role both as a person and as a creature of my species; and a looming existential dread that sometimes attains the physicality of a lump in the throat, a knot in the abdomen, a constriction around the temples. They seem important, these various elements of response, like basic conditions of existence. (1999, 13–14)

And then there is theologian Douglas E. Christie, who is familiar with that same Mattole River from time spent in a monastic community inside the Mattole watershed. He offers this following reflection:

An encounter with salmon, or any living species, or the entire fabric of life, invites us to ... open ourselves and respond to the mysterious Other with honesty and imagination and, perhaps, faith. Learning to live with such openness toward the Other, to become immersed in the "vast complexity" of the world is profoundly erotic, contemplative work, and a fundamental part of our learning to inhabit the world and take responsibility for it. (2013, 262–3)

If it is true that the dominant narrative of human dominance is largely self-proclaimed, then it would also be true that the narrative can perpetuate itself only for as long as a critical mass accepts its claim to legitimacy. But that legitimacy is being eroded every time salmon breach the surface of appearances and ignite inside one of us a sense of wonder, or beauty, or compassion. It is eroded every time one of us finds her thinking mind not pitted above or against but in an eddying confluence with the life of the body and the larger life of rivers, mountains, or forests.

To the degree that we encourage one another to pay heed to such wordless exchanges, and that we create the conditions for such experiences to flow forth in the first place, the once-dominant narrative will no longer seem so inevitable or commonsensical. Instead, we might find that its legitimacy is already waning, while a different kind of critical mass is surging up against its eroding foundation: it is that more-than-human alliance to which many are pledging their allegiance today, lending to it not only their best intellectual capabilities, but also their vulnerability, their grief, their deep joy, their creativity, their care, their sense of beauty.

And sure enough, this already has direct, concrete impacts to show for itself. Here is Lynda Mapes (2016) again, reporting from the Elwha early in 2016:

Elk stroll where there used to be reservoirs. Bigger, fatter birds are bearing more young, and moving in to stay. A young forest grows where there was blowing sand in the former reservoir lakebeds. Seeds tumble down the river's coursing current. The big pulse of sediment trapped behind the dams is passed; the river has regained its luminous teal green color, and its channel is stabilizing.

Logs are tumbling and stacking in the river, building complex, braided channels, islands and jams.

And fish are booming back: More than 4,000 chinook spawners were counted above the former Elwha Dam the first season after it came down. Overall, fish populations are the highest in 30 years. And that's before the first progeny of salmon and steelhead going to sea since dam removal come back this year.

For metamorphic fish whose ancestors may have continuously lived inside this land and its streams for as much as fifty thousand times a thousand years (Lichatowich 1999; Montgomery 2003; Coates 2006), and who have endured despite repeated glacial advances and changing climates, what, really, are concrete obstructions whose life spans are measured in centuries at best? For a century, the Elwha salmon could not return, but what they could do was continue to insist that they would not give up trying. And they could continue to call out to those who had it in their power to dismantle the dams. Until, at long last, a critical mass of people finally understood what the salmon had been saying all along. It may not be too much to say that the salmon have played an active part in the dismantling of the two Elwha dams, and in the larger revision of values and desires underway throughout the region.

Life Wants to Live

It bears repeating: Voices that continue thinking of salmon as inconvenient disturbances to industrial development are *not* uttering inalienable truths; their claims to legitimacy are *not* unchallengeable. They may – while conflicts still flare up – co-opt such notions as sustainability or even responsibility, but they cannot, once and for all, contain the persistent upwelling of wonder in the encounter with wildness, or blockade the spawning, sprouting, birthing, and hatching of new life, or obstruct the instant and intuitive recognition of kinship between fly fisher and salmon, or seal the countless ways in which our breathing bodies still respond alertly, and competently, to the voices of river, wind, or estuary. Our mindful bodies are still being drawn towards, called upon, awakened, stirred, and roused by rainstorm, solstice, or autumn moon, by moose or beaver or wolf; still, salmon radiate a particularly vigorous eloquence and enflame a special kind of awe in us, charging encounters between our kinds, now as ever, with that profoundly erotic tension of which Christie writes. These are dynamics worth taking seriously, for that which keeps surging and leaping and running up against the physical and metaphysical dams of the human-centred lifeworld is none other than life itself, raucous, untamable life, wanting to live.[1] This may be warning or pledge, depending on where our allegiances lie: Life will not be contained or owned. Really, it never has been.

NOTE

1 The California-based philosopher Derrick Jensen first coined and explored this beautiful phrase, "Life wants to live," in *Dreams* (2011).

REFERENCES

Abram, David. 2010. *Becoming Animal: An Earthly Cosmology.* New York: Vintage.

Aldwell, Thomas. 1950. *Conquering the Last Frontier.* Seattle: Artcraft Engraving and Electrolyte Company.

Cermaq. 2014. "Salmon Farming in BC Has Grown into a World-Class, Highly-Advanced Industry." Accessed 10 August 2016. https://www.cermaq.com/wps/wcm/connect/msca-content-en/mainstream-canada/aquaculture/salmon-industry/.

Christie, Douglas E. 2013. *The Blue Sapphire of the Mind: Notes for a Contemplative Ecology.* New York: Oxford University Press.

Clayoquot Action. 2015. Accessed 25 February 2018. http://clayoquotaction.org/.

Coates, Peter. 2006. *Salmon.* Chicago: University of Chicago Press.

Colombi, Benedict J., and James F. Brooks, eds. 2012. *Keystone Nations: Indigenous Peoples and Salmon across the North Pacific.* Santa Fe: School for Advanced Research Press.

Crane, Jeff. 2011. *Finding the River: An Environmental History of the Elwha.* Portland: Oregon State University Press.

Davis, Marc. 2016. "Foreign-Owned Fish Farms Are Devastating B.C.'s Wild Salmon." *Huffpost British Columbia.* 03 February 2016. Accessed 25 February 2018.http://www.huffingtonpost.ca/marc-davis-/fish-farming-wild-salmon_b_9361814.html/.

Deloria, Vine Jr. 2012. *Indians of the Pacific Northwest: From the Coming of the White Man to the Present Day.* Golden, CO: Fulcrum Publishing.

Duncan, David James. 2001. *My Story as Told by Water.* San Francisco: Sierra Books.

FDA. 2015. *FDA Has Determined that the AquAdvantage Salmon Is as Safe to Eat as Non-GE Salmon.* Consumer Health Information. Accessed 20 August 2016. http://www.fda.gov/downloads/ForConsumers/ConsumerUpdates/UCM473578.pdf/.

Hannesson, Rögnvaldur. 2010. "Hva skal vi med villaksen?" *Dagens Næringsliv* 26.

House, Freeman. 1999. *Totem Salmon.* Boston: Beacon Press.

Jensen, Derrick. 2004. *Endgame.* New York: Seven Stories Press.

– 2011. *Dreams.* New York: Seven Stories Press.

Johnsen, D. Bruce. 2009. "Salmon, Science, and Reciprocity on the Northwest Coast." *Ecology and Society* 14 (2): 43. Accessed 25 February 2018. https://doi.org/10.5751/es-03107-140243. http://www.ecologyandsociety.org/vol14/iss2/art43/.

Kristoffersen, Svein. 2011. "Vil satse bærekraftig." *Klassekampen* 9. 18 February.

Lichatowich, Jim. 1999. *Salmon without Rivers: A History of the Pacific Salmon Crisis.* Washington, DC: Island Press.

Manning, Richard. 2004. "Life's Deep Kinship." *Patagonia.* Accessed 24 August 2016. https://eu.patagonia.com/enGB/patagonia.go?assetid=9100/.

Mapes, Lynda. 2013. *Elwha: A River Reborn.* Seattle: The Seattle Press.

– 2016. "Elwha: Roaring back to Life." Accessed 25 February 2018. http://projects.seattletimes.com/2016/elwha/.

Montgomery, David. 2003. *King of Fish: The Thousand-Year Run of Salmon.* Cambridge, MA: Westview Press.

Morton, Alexandra. 2016. Blog. Accessed 25 February 2018. http://alexandramorton.typepad.com/.

Mueller, Martin Lee. 2016. *Being Salmon, Being Human: A Phenomenology of Story.* PhD dissertation. Oslo: University of Oslo.

Nelson, Richard. 1991. *The Island Within.* New York: Vintage Books.

The Norwegian Government. 2015. *Meld. St. 16 (2014–2015). Forutsigbar og miljømessig bærekraftig vekst i norsk lakse-og ørretoppdrett.* Oslo: Det Kongelige Nærings-og Fiskeridepartementet.

Waldau, Paul, and Kimberly Patton, eds. 2006. *A Communion of Subjects: Animals in Religion, Science & Ethics.* New York: Columbia University Press.

Woelffe-Erskine, Cleo, July Oskar Cole, and Annie Danger, eds. 2007. *Dam Nation: Dispatches from the Water Underground.* New York: Soft Skull Press.

Wolf, Edward C., and Seth Zuckerman, eds. 2003. *Salmon Nation: People, Fish, and Our Common Home.* Portland, OR: Ecotrust.

PART TWO

Water and Place

While our initial tendency may be to visualize "place" as geographical location, we must not lose sight of the fact that physical spaces function as more an mere material containers of human activities. To *be* is always to *be somewhere*. The fact is that places "serve not merely as the settings for our lives, but as participants, as vibrant, living aspects of memory, tradition, history, and meaning" (Donohoe 2014). Our moods, awareness, where we are situated culturally, historically, spatially – define each of us, constituting an interpretive horizon of intelligibility of the world in which we find ourselves. In the words of Noel Arnaud, "I am the space where I am" (cited in Bachelard 1964, 137). Indeed, for this reason, human existence itself is, properly speaking, defined as *implaced* (Heidegger 1971; Casey 1993).

In many ways, one can argue that the current debate about the significance of place emerged from the classic writings of German philosopher, Martin Heidegger. Meditating on the meaning of building as dwelling, he pointed out that dwelling itself "is the manner in which mortals are on the earth ... We do not dwell because we have built, but we build ... because we are *dwellers*" (1971, 148). In the words of place guru, David Seamon, "there is no dualistic person/world or people/environment relationship. Instead, there is only a people-world immersion, entwinement and comingling ... person-intertwined-with-world" (2013, 144). This fundamental, ontological belonging of human existence and place is the beginning of any philosophical discussion of how meaning emerges in decisions that we make around public policies and planning frameworks.

Moreover, our dwelling places should be seen as more than simply the consequence of human construction or engineering of the natural and built environments. Rather than a product of our making, dwelling has the character of humility and care, of sparing and preserving.

"Real sparing is something *positive* and takes place when we leave something beforehand in its own nature, when we return it specifically to its being" (Heidegger 1971, 149). Heidegger even goes so far as to call such dwelling poetic: "Poetry and dwelling belong together, each calling for the other" (227). His writings, while unorthodox, serve to remind us to conceive of policy making, planning, and design as less a consequence of merely technical manipulation and human projection and more a matter of building a creative sensibility and finding the right *fit* between our needs and the world that sustains us.

Just as we cannot *be* in the absence of place, so too is that implacement defined through our existential belonging to water. Diverse means of implacement and engagement with our world are shaped by our relationships to water. Whether its absence emerges through moments of drought or through its buried meanderings that shape our landscapes – or whether its abundance overwhelms us in floods or in the glory of a seascape, water is integral to our sense of place and our understanding of who we are.

This section begins with **Janet Donahoe's** reflections on the place of water as essential to all human dwelling. This chapter introduces the significance of place as a philosophical concept while also demonstrating how placemaking informed by a sensibility to water results in better planning of our built environments.

Irene Klaver's magnificent chapter then takes us to a discussion of "monstrous" waters that meander throughout our lived environments, sometimes as deadly childhood places and other times as inequitable, undemocratic gentrified urban spaces. Advocating for a "riverspheres" approach to city planning, Klaver shows how simplistic, geometrical models of design and thought need to be replaced by "models of complexity, of flows; not just flows of water but of people, capital, light, luggage, tourists, money, exchange and experiences."

Ingrid Leman Stefanovic explores how natural cities are defined by the place of water. Thoughtful architectural design and planning practices provide opportunities for building our human settlements in a way that both respects and celebrates water's place within our lived experiences.

Sarah King then illustrates how the poisoning of one city's water system is only one symptom of deeper cultural paradigms of racism and power inequities that result in marginalized communities and unethical practices.

King's chapter nicely transitions us through her conversation about lived inequities to Part Three, where we begin to delve more deeply into how to rethink the ethical and lived dimensions of current policies and practices about water.

REFERENCES

Bachelard, Gaston. 1964. *The Poetics of Space*, Translated by Maria Jolas. Boston: Beacon Press.

Casey, Edward. 1993. *Getting Back into Place: Toward a Renewed Understanding of the Place-World*. Bloomington and Indianapolis: Indiana University Press.

Donohoe, Janet. 2014. *Remembering Places: A Phenomenological Study of the Relationship between Memory and Place*. Lanham, Boulder, New York, and London: Lexington Books.

Heidegger, Martin. 1971. "Building Dwelling Thinking" and "Poetically Man Dwells," in *Poetry, Language, Thought*, translated by Albert Hofstadter. New York: Harper & Row, Publishers.

Seamon, David. September 2013. "Lived Bodies, Place and Phenomenology: Implications for Human Rights and Environmental Justice." *Journal of Human Rights and the Environment*, 4, no. 2, 143–66. https://doi.org/10.4337/jhre.2013.02.02.

5 The Place of Water

JANET DONOHOE

Two decades ago, Edward Casey enlivened our thinking about place and reinvigorated our attempts to come to an understanding of the role of place in our lives. In those discussions, Casey (1993) introduced a way of thinking about the distinction between orientation and dwelling. To capture that distinction, he writes of being asea on a vessel. The sea, for those of us who are not seafarers, is a vast space of indeterminacy requiring us to use various tools in determining our whereabouts. We must orient ourselves in this vast space. How different that orientation is from our impetus to dwell upon land within towns and cities that we make familiar through repetition and by marking the landscape with aptly named landmarks.

While such landmarks provide for our orientation, they do much more than that. They allow us to dwell. It is compelling that Casey uses the sea as the foil for our dwelling, for it raises the question of the role of water to our dwelling. While it might be clear that we generally do not dwell on the sea, this does not mean that we do not dwell with water. In fact, water plays a vital and particularly unexplored role in our dwelling. Thus, in the following, I want to think through that role of water to dwelling, which necessarily means thinking through the role of water to place, in place, and for place. I will begin with an explication of the notion of dwelling, and follow that by elaborating on many of the ways in which water is constructive of place and conducive to place. However, I will also examine ways in which water is destructive of place to complicate our thinking about water and place. Fundamentally, I will show that water is most proper to our dwelling in its reflection of the human condition.

Dwelling

Casey's introduction of the notion of dwelling is presented in his *Getting Back into Place* (1993, 109). He draws heavily on Heidegger's conception of dwelling as set out in "Building, Dwelling, Thinking" (1977a). In that text, Heidegger associates dwelling with our being-alongside in the way that we *are* on this Earth. For Heidegger, dwelling is fundamental to our being. As the title of Heidegger's essay implies, dwelling is associated with building. To dwell, we must build. In building, we provide a middle ground upon which we can rest, a place to fend off the harshness of nature and foster our peculiarly human activities. Yet Heidegger recognizes that not all buildings are dwelling places. Nonetheless, things we build contribute to our dwelling. As he writes, "Those buildings that are not dwelling places remain in turn determined by dwelling insofar as they serve man's dwelling. Thus dwelling would in any case be the end that presides over all building ... to build is in itself already to dwell" (321). Buildings, whether they are structures in which we dwell as homes, or structures in which we work or play, physically provide for our dwelling. At the same time, they serve a more symbolic purpose in fending off the constant change of human existence by providing places of stability. Casey elaborates: "A building is also a compromise formation at another level. Thanks to such features as stability and enclosure, it arrests accelerated movement and allows the lived body to rest. If the same body reenters the open world, it often seeks to return to the habitation from which it set out" (1993, 112).

To provide that stability, most built places are constructed to last for a period of time. Perhaps that time is relatively short, as when we pitch a tent, but many are for periods much longer than that – when, for instance, we build a brick home or a stone monument intended to last for generations. In spite of the differences of these dwelling places, we can identify important commonalities. In fact, Casey recognizes two characteristics that we aim for in our building: 1) buildings must be constructed for repeated return, and 2) they must possess a felt familiarity. Dwelling places with these characteristics allow us to move beyond orientation to inhabitation. Inhabitation involves bodily familiarity that makes a place feel homey to us. We become accustomed to a place and are capable of dwelling with(in) it. The importance of this aspect of dwelling is that it provides stability, a kind of permanence for our embodiedness. This struggle for stability and pseudo-permanence marks the human endeavour. And this endeavour primarily takes place on land, where we can orient ourselves and where we can build. In built places, then, we have both orientation and inhabitation.

In contradistinction to our finding our orientation on the ocean, Casey sees places of familiarity and repeated return as places in which we can dwell. He claims that "being lost at sea ... means lacking place in an endless space-world. Yet we can even have exact bearings in such a world and still not be fully placed. Our knowledge of local seamarks may be intact and modern navigation may pinpoint our position in world-space, but in terms of implacement something remains missing" (1993, 109). In that vast expanse of water, something is missing, something that makes it impossible to truly dwell. Casey argues that the missing elements are the stability and inhabitancy that compose dwelling. These are lacking in the space of the sea. Casey further draws upon Heidegger's "Building, Dwelling, Thinking" to argue that dwelling depends upon cultivating place to meet these conditions of stability and inhabitancy.

By cultivation, Heidegger means two things: caring for the fields in the sense of cultivating crops, and caring for ourselves in the sense of creating culture. I would like to suggest that although the dwelling described so far is not something that can be done on the water, both kinds of caring for involve or are related to water. Of course, to cultivate fields and grow crops, we must have water. And frequently we bring water to the crops through irrigation or even a watering can. But, culturally, we engage with water in our caring for as well. As we build culture, we build ways to manage water, to control its movements through city streets to reduce its erosive effect or avoid its pooling in dangerous ways. Aqueducts are one of the surest signs we have of culture. Caring for, then, has in large measure become a matter of managing, controlling, and manipulating water. These different forms of caring for reveal our ambiguous relationship to water. While we see water in its life-giving necessity, we also have deep-seated anxiety about water, not as life-giving, but as alienating and even death-bringing. Just as we hope to control, manage, and ultimately thwart death, so too we desire to control and manage water. This is our first hint of the ambiguous relationship between water and place.

We have begun, then, with a dichotomy between the constant change of the high seas and the stability of terrestrial places. Terrestrial spaces that are not places include, like the sea, spaces that are expansive and resist our ability to orient ourselves, such as open prairie or large tracts of forest. In such spaces, we are easily lost, unable to get our bearings. What this means about places is that they serve to demarcate and contain, to some degree. Frequently, water also serves this purpose of demarcating and creating borders. Rivers separate while larger bodies of water create limits for possible growth of cities. Places have edges and borders, and can be differentiated from surroundings. Places shelter or contain our lives, thereby allowing us, even encouraging us, to dwell.

This dichotomy between places and spaces is somewhat false, however, since it trades on the notion of the stability of built places. For we can surely say that with our body as our primary point of location, the stability of any place is only temporary. Our body changes, places change, our relationship to places changes. Moreover, the stability that we seek is called into question by the very presence of water within place. The movement of water – its journeying, as Heidegger would say – puts us in mind of the constant change that is the human condition. Stability is illusory. We provide what kinds of it we can, but such provisions are merely temporary and are only significant insofar as we sense our own journey in spite of such attempts. So, while the experience of place may be primary, it is shot through with the anxiety of no place, the anxiety of changing and changeable space, water.

Water, Home, and Journeying

We build with water in mind, both in the sense of aiming to keep water out by having a roof over our heads and in terms of incorporating water as a convenience. We do not build *with* water. We can and do incorporate water into what we build, but for it to be incorporated, it must be contained, channelled, or controlled in some way. Yet, we are eager to include water in spite of the difficulties of doing so. Water contributes to our dwelling in ways that perhaps have more to do with cultivation than building. We wash our clothes in the river, scatter ashes of the dead there, play there, travel there, bathe there, acquire food there, give birth there. Without water, there is no culture. We use it in all manner of places to create an atmosphere of calm: for sacred purposes, for exotic purposes, for purposes of reflection both literal and spiritual. It calls us back to ourselves in a place. Water is inseparable from our dwelling.

   Water can also be spectacle, as when we hike to see a waterfall. We seek it out on such hikes, we are drawn to it and we face it at the beach rather than having it at our backs. Great cities across the globe such as New York, Chicago, Amsterdam, Venice, Stockholm, Detroit, St Petersburg, and Bangkok are built around water. These cities incorporate water not only out of necessity or convenience; they are not just port cities, but are cities that have incorporated water by design. Water has contributed to making some of these cities the greatest in the world. Water has been deemed necessary to dwelling.

   We gather around water as a source of life. When we encounter water outside of built places, it can be a place in and of itself. We dwell in and through it. It seems neither cultivated nor built, yet contributes in significant ways to our dwelling. Heidegger describes this with respect to the river Ister – the referent of Hölderlin's poem "The Hymn to the Ister," which Heidegger analyses extensively. In Heidegger's analysis, he suggests that "we also learn that the rivers are a distinctive and significant locale at which human beings, though not only human beings, find their dwelling place" (1996, 12). The river, of course, can be a place. We mark rivers on maps. They serve as landmarks. They frequently serve as borders. They are places we gather, places we go for recreation, places we traverse for commerce, places we build. As Heidegger contends, "the river determines the dwelling place of human beings upon the earth" (20).

But for Heidegger, the river is more than that. The river is associated with home – the home of Earth, but also the home in a more essential way of what is necessary to our living. As he remarks, "[t]he river *'is'* the locality that pervades the abode of human beings upon the earth, determines them to where they belong and where they are homely [*heimisch*]. The river thus brings human beings into their own and maintains them in what is their own" (21). What Heidegger means by this is that the river in its journey is what inclines humans to build, to seek the stability, repeatability, and inhabitancy of a home. The river is journeying, never stopping to rest, never ceasing its movement. This makes the river a constant of change, as was remarked upon by Heraclitus. Heidegger reinforces this notion by claiming that "[t]he river is simultaneously vanishing and full of intimation in a double sense. What is proper to the river is thus the essential fullness of a journey. The river is a journey in a singular and consummate way" (30). And further, "This journeying that the river itself *is* determines the way in which human beings come to be at home upon this earth" (30). The journeying that is fundamental, even essential, to rivers is what inclines humans to dwelling, thus to building, to create upon this Earth a place for ourselves and a home in which to dwell. Finally, Heidegger contends that "[t]he journeying that the river *is* prevails, and does so essentially, in its vocation of attaining the earth as the 'ground' of the homely" (30).

In the broadest sense, the place of our home that establishes us as Earthlings is itself a place of water. More than 70 per cent of the Earth's surface is covered with water. Our home place, the very place where we ourselves have become what we are, is characterized by water. Even in our efforts to escape this place, we know that what we must seek elsewhere to make any other place a place of our home is water. Our very dwelling, then, is marked by the tension between the journeying and the seeking of stability and inhabitancy. This tension is not one to overlook or attempt to escape; it is the very characteristic of our dwelling. Perhaps one reason that so many dwelling places incorporate water is due to the need or desire to hold the journeying in tension with the stability. For we really cannot need the stability without the constant threat of change. It is a reminder of our impermanence on this Earth. We build to stave off that impermanence, all the while knowing that the waters will come, that change is inevitable. Dwelling is temporary. To hold the stability and the threat of change in tension is the horizon of human existence.

For in our dwelling, we are historical beings, never ceasing to be on a journey while at the same time cultivating and caring through building and establishing culture. As Heidegger notes,

> [b]ecoming homely and dwelling upon the earth are of another essence. We may approach it in giving thought to the essence of the rivers. The river is the locality for dwelling. The river is the journeying of becoming homely. To put it more clearly: the river is that very locality that is attained in and through the journeying. (31)

Journeying itself is between places. As we journey, we move from one place to another. Places are connected by the journey. Rivers, likewise, can be the path of our journey as we move along the river or across the sea from port to port. The water in its flow connects us to place and connects places to each other. Sometimes the river makes journeying possible. As Remmon Barbaza puts it:

> We know that the history of civilization is always closely tied with water. Nearly every great city was established near rivers or by the sea. And in any case human life is not possible in places where there is no freshwater or no possibility of accessing it from a neighboring place. This historical fact alone – that we build our lives around water – manifests our dependence as human beings and communities on water, particularly freshwater. (2012, 64)

Our dependence upon water has the effect of our wanting to incorporate water into our cities and homes. This cultivation of water, however, is also ambiguous.

> We have, perhaps, taken our cultivation of water to a different extreme in linking it too closely to dwelling, such that we have attempted to domesticate water (see Macauley 2010, esp. ch. 7). David Macauley has commented on the domestication of water in such places as Frank Lloyd Wright's "Fallingwater," where we "see the congress of stability and habitability of the architecture on the one hand with the flux of the moving fluid on the other hand," allowing us to see the "captured moment and continuous momentum of domestication in which the meaning of the

stream is suddenly and spectacularly redefined in its arranged marriage to Wright's home. (262)

Domestication of water does not stop with the domicile, however. It includes the vast variety of ways in which we attempt to control the water in our midst without paying attention to its essence as river, lake, pond, rain, snow, ice cap, or sea. Our cultivation of water entails aesthetic uses as well – water elements of monuments and memorials come to mind. In such cases, we manipulate water through pipes to get it to bubble or fall just so for aesthetic appeal or symbolic meaning. We use water in such instances for some kind of dramatic effect.

As Heidegger taught us in "The Question Concerning Technology" (1977b), modern thinking, like the power plant on the Rhine, appropriates water and attempts to store it as energy, creating of it a standing reserve. The standing reserve is there for our use and our convenience, thereby losing its own manner of being. In the case of places like the Hoover Dam or the Aswan Dam, we attempt to cease the river's journey until we deem it appropriate for the water to continue on its way. We attempt to control that journey. Moreover, the move to incorporate water into urban landscapes treats water as entertainment – instrumentally for the sake of making money. We see this in the paving over of rivers, the establishment of water parks, the management of release of water from dams for the sake of recreation, and even the popularity of riverwalks in many cities that establish a touristy, commercial area along the banks of a river. We also see it in an extreme way in the use of water for the extraction of sources of energy from the land in fracking, coal, and oil extraction. In such cases, we use the standing reserve of water to create a standing reserve of coal and oil that we rip from the Earth. As Barbaza states, "we have reversed this relation of dependence, making water adjust to our lives and projects and ambitions, as when residential communities built on watersheds or huge dams erected on rivers force water to redirect itself as it seeks its own level, and many times with deadly repercussions" (2012, 64). The deadly repercussions of our instrumental relationship to water are many, not least of which is the contamination of drinking water through these energy extraction processes.

An instrumental relationship to water, at some level, cannot be avoided, since human life is impossible without it. Where there is a scarcity of fresh water, there are few humans. So, to some degree by necessity, we build near water. At the same time, however, our approach to water has exceeded this respectful necessity of instrumentality towards a simple use-factor. Due to the domestication of water, we have forgotten our relationship to it. We in the developed world take it for granted, not recognizing the ways in which we treat it primarily instrumentally.

At the same time, we can ask whether the domestication of water is fundamental and necessary to dwelling. Of course, the answer must be no. Our cultivation, which Heidegger deems essential to dwelling, is not about the application of technology to water, or the management of water. It is about the ability to care for water and to take care with regard to water. In doing so, we can find ourselves more attentive to water in its waterness, thereby bringing our own placemaking in line with water rather than bringing water in line with our placemaking. To respect water in allowing it to reveal itself as what it is means that the place of water must be (no pun intended) fluid. As Barbaza (2012) reminds us,

> To a large extent of course the goals of better water management and governance are necessary and the efforts made towards it laudable. But we might do well to make room for the question whether an approach that is primarily managerial and technical is enough to make us realize the kind of stance we need to adopt in relation to water, or to know whether we need to restore a primordial relationship with water that we lost in the midst of modern technology. (65)

In other words, we question whether a managerial approach to water will solve the water problems that have arisen as a result of our managerial approach to water. Perhaps we need to understand our ambiguous relationship to water in its complexity.

From what we have described above, our primordial relationship to water is ambiguous. Water, in its journeying, is fundamental to our building and dwelling. It is essential to our homemaking upon this Earth and of this Earth. At the same time, our contemporary efforts to control and manage water limit our abilities to recognize water in its essential relationship. We take up water instrumentally and rarely allow it to be as it is in its journeying. The results of our inability or unwillingness to understand that essential and essentially ambiguous relationship to water means that we will be increasingly confronted by water's destruction of our place and places on this Earth.

Water and Destruction of Place

While we are certainly eager to say that water contributes to our dwelling and that we could not dwell without it, the ambiguity of our relationship to water gives us reason to be wary. For water can frequently be devastating to our dwelling. In this final section, I will suggest that even in those instances when we think that water is destructive of place, we might be able to see that the destruction is sometimes necessary or conducive to dwelling. Although Aristotle remarks that "place is not destroyed when what is in it is destroyed" (1957), our capacity for dwelling in a place can be radically altered. And perhaps our capacity to dwell *with* the Earth rather than *upon* the Earth will be enhanced.

It is no surprise that water can powerfully destroy places. Tsunamis, hurricanes, floods, erosion, and ice melts are all water events with which we are familiar – if not through personal experience, then at least through news and media images. Catastrophic and tragic water events have become increasingly violent and destructive in the past few decades, as the Earth's temperatures and thus the ocean's water levels begin to rise. It is not new to see water in its destructive form, however. Flood stories from as far back as the writing of the Bible and the *Epic of Gilgamesh* raise the spectre of the destructive force of water. When we are personally confronted with such destructive force, we are overawed, even afraid. It is a powerful force that brings to mind our mortality, if not serving up mortality on a massive scale. Water has the capacity to erase us and our places from the face of the Earth. It can cleanse the Earth of our presence and restore the Earth to itself. While we may be able to manage or control water for a time, in these raging forms, water can rarely be contained, controlled, or mastered.

What happens when water destroys terrestrial places? Can place truly be erased by water, replaced by water? Consider what happened to villages that were submerged when the Aswan Dam was built. On the border between Egypt and Sudan, the dam's construction was begun in 1960 and not completed until 1968. The dam was built to control the annual flooding of the Nile River, and to provide electrical power and irrigation to much of Egypt and Sudan. Also, the dam has made navigating the river easier, allowing for increased tourism, thereby having a commercial impact on the region. The dam created Lake Nasser, which covered over part of Nubia, displacing about 100,000 people. Their homeland was forever destroyed, while they were relocated to other parts of Egypt and Sudan. An ancient Egyptian temple complex had to be relocated rather than be submerged.

The Aswan Dam project exemplifies the destruction of a place through an attempt to harness or control the water. The water, then, becomes its own place, Lake Nasser, site of a thriving fishing industry. This new place has replaced the homes of those forced to leave. Their farming culture that used the silt from the annual floods of the Nile to fertilize their fields and to make them highly productive has been replaced by the artificially controlled waters that are devoid of silt; it gets trapped in the dam, leading to the use of millions of tons of artificial fertilizer for far less productive crops. The damming of the Nile has also destroyed or caused the relocation of many archeological sites, meaning that this place of ancient cultures is now completely covered over and inaccessible.

In the attempt to control the Nile, we are confronted with both the life and the death that coexist in the place of Lake Nasser. It has destroyed the villages and the way of life of the farmers who used to live there, while serving the more highly instrumental approach to the water as standing reserve. One form of caring for has usurped another. One kind of place has been replaced with another.

Another example of the destruction of place by water is New Orleans. The waters that were to have been controlled by levees that had already altered the place ran rampant as a result of Hurricane Katrina in 2005, overflowing the controls, overwhelming the mastery to reassert themselves as river, lake, and gulf coastal area. The destruction of New Orleans was not, of course, as complete as that created by the Aswan Dam, and yet, the effects of the water would completely reorient the city. What was initially a physical marker of the uniqueness of New Orleans remains just that, although in a completely altered way and for completely different reasons.

When we consider these examples, we might initially be inclined to distinguish between the creation of Lake Nasser and the flood of New Orleans. After all, one is a humanly created event, while the other we call a natural disaster. Upon closer inspection, however, we may begin to understand that they are not so different. They are considered disasters only from our human perspective. Had we not built so close to the water, had we not contributed to climate change, this destruction of place could perhaps have been averted. But we cannot seem to stop ourselves. Rather than letting the waters be as they are, we attempt to control and manage, to butt up against them in our hubris and in our confidence in our technologies. Of course, we must be careful when we consider this argument about managing water through technology. In such arguments, we have a tendency to identify technology as something recent, within our lifetime. We conceive of it as new. But technology, particularly as it concerns water, goes back to the first attempt to channel or mitigate the movement of water. We immediately think of the aqueducts of ancient Rome, for example. So, we must be careful to recognize that the pipes that bring water into our homes are also forms of technology and control about which we must think carefully. Along with Heidegger, then, we must stress that we are not condemning water technology or the domestication of water in all its forms. Rather, we are drawing attention to the need to reconsider the ways in which place and water are interconnected and how water is used for place or in place.

Conclusion

Fundamentally, we must step back and look at what our dwelling with water and our incorporation of water into places is all about. What I have shown here is that our relationship to water in place is ambiguous. We frequently suggest that water is so important because it gives life. Its position is sacred due to its strong connection to life. But what I have underscored here is that we can also see our connection to water in place as a reflection of our human, all too human, confrontation with death. The power of water lies in its ambiguity, in both giving and taking life. As a reminder of our indeterminate time, of the ephemerality of our living, of the fragility and power of ourselves, water is both creator and destroyer, both life-giving and life-taking, healing and killing, soothing and anxiety producing. In this way, water is the most proper reflection of the human condition.

At the same time, we must recognize that our penchant to control and even alter that human condition spurs us to attempt to control and manage water. To recognize what it means to dwell with water is to see that our instrumental approach to water has its dangers. So many of us are not fortunate enough to be able to dwell with water, let alone have access to clean, life-sustaining water. And so many others build too close to coastlines or on flood plains where water can destroy. Others seem intent on persisting in activities that we know will raise the levels of the oceans, threatening even more places with destruction. To be mindful of the place of water will, of course, not solve these issues, but it might bring us a step closer to thinking more clearly about how water figures in our dwelling.

Of the four elements, water is the only one that literally has the power to reflect us back to ourselves. And when it does so, we must ask ourselves what we see. We need to recognize the importance of water to place and human dwelling. We cannot disregard its role in showing us who we are and calling us to account. With respect to water, it seems we may still be at sea.

REFERENCES

Aristotle. 1957. *Physics.* Translated by P.H. Wicksteed and F.M. Cornford. Cambridge: Harvard University Press, 209a1.
Barbaza, Remmon. 2012. "Letting It Flow: Towards a Phenomenology of Water in the Age of Modern Technology." *Kritika Kultura* 18: 64. https://doi.org /10.13185/kk2012.01804.

Casey, Edward. 1993. *Getting Back into Place*. Bloomington, IN: Indiana University Press.

Heidegger, Martin. 1977a. "Building Dwelling Thinking." In *Basic Writings*, edited by David Farrell Krell. New York: Harper & Row, 319–39.

– 1977b. "The Question Concerning Technology." In *Basic Writings*, edited by David Farrell Krell. New York: Harper & Row, 284–317.

– 1996. *Hölderlin's Hymn "The Ister."* Translated by William McNeill and Julia Davis. Bloomington, IN: Indiana University Press, 12.

Macauley, David. 2010. *Elemental Philosophy: Earth, Air, Fire, and Water as Environmental Ideas*. Albany, NY: SUNY Press.

6 Engaging the Water Monster of Amsterdam: Meandering Towards a Fair Urban Riversphere

IRENE J. KLAVER

Bullebak

The day I was born, they baptized me. The priest poured water over my head and initiated me into the Catholic Church with the Dutch version of the words "In the name of the Father, and of the Son, and of the Holy Spirit." I probably cried loudly. Most babies cry at their baptisms. Maybe we are supposed to cry, expressing a bodily participation in the cultural initiation. My loud tears performed a first mingling of my own watery fluids with the ritualistic waters – in this case, the waters of a supposedly holy sacrament washing away an original sin. It would have been different if I had been born elsewhere. I don't remember anything of it. I don't remember the warm waters of the womb. I don't remember the cold waters of initiation into a patriarchal religion. I do remember the Bullebak.

We lived in the last house of the village Opmeer, next to a watery ditch that was as green as the grass of the pastures on the other side. In that part of North Holland – West Friesland, north of Amsterdam – the soil used to be rather wet. The land was low and the groundwater level high. Especially the polder areas, "conquered" from the sea or inland water bodies, needed to be drained continually by an elaborate system of pumps, dikes, and ditches. We wore wooden shoes on those lands. With regular shoes, you got stuck in the mud. Wooden shoes functioned as small boats on the soppy clay substratum. If you squinted at the landscape, you could barely distinguish the green duckweed-filled ditches from the grasslands. The ditches were dangerous for kids. They were deep enough for us to drown. Our mother told us to stay away from them: the Bullebak lived there, she said.

The Bullebak had big hands and strong arms; he was also part fish, big, with sharp teeth. He was a water monster. If we came too close to his realm, the water, he would grab us and take us with him. Forever.

When I turned four, I joined the motley troupe of brothers, sisters, nieces, nephews, and neighbour kids walking to the school in the next village, Hoogwoud. Trees lined the straight country road, and a ditch separated the road from the pasturelands. The art was to walk behind the trees without falling into the ditch. Mom knew our game, our obsession. She knew we would not stay on the road side of the trees, but were drawn to the dark danger of the ditch. "*Kijk uit voor de Bullebak!*" she emphasized every day when we left for school, trying to instill fear of the water in us: "Beware of the Bullebak!"

About a year later, we moved to the town of Alkmaar, 70 kilometres north of Amsterdam. We still walked to school. There were no ditches along the road, just other streets and traffic. Now the art was to not step on the lines between the sidewalk tiles, to lie on the freshly cut hedge of the stately house without being caught, and so forth. Mom did not mention the Bullebak anymore. Now she emphasized, every day when we left for school, to look left and right and left again before we crossed the street, and not to talk to strangers, especially not to men. The Bullebak was left in the ditches of Opmeer, our previous life.

For years, I never thought about the Bullebak: not until the mid-1980s, when I was a university student and lived in the centre of Amsterdam. One day I happened to bike over the Brouwersgracht to the Marnixstraat. I was about to cross the bridge into a poorer part of town when I noticed its name – in the characteristically capital-letter style of many Amsterdam bridges – BULLEBAK (see Figure 6.1). I almost fell off my bike. The Bullebak was more than family lore! He was not just my mom's invention, a cautionary tale told to keep us away from the water. The Bullebak loomed in a larger cultural imagination. Although, as it turned out, not that large. College friends from other parts of the Netherlands than West-Friesland seemed not to have heard of him.

The Brouwersgracht, with the BULLEBAK Bridge, forms the northern border of the so-called *grachtengordel*, the Canal Belt, connecting the major canals – Singel, Herengracht, Keizergracht, and Prinsengracht – to the outer Singelgracht. In the seventeenth century, the Golden Age of the Netherlands, the seafaring Dutch had become world leaders in trade, science, industry, and arts. The monopoly on Asian trade of the Dutch East India Company brought silk and spices from the Far East to Amsterdam. The warehouses at the Brouwersgracht stored the valuable goods from the large sailing ships. From there the goods were distributed all over Europe, and all over the world. By the twentieth century, the ways of commerce and the place of The Netherlands in world history had changed dramatically. With the Dutch having lost their central economic position, most depots on the Brouwersgracht had lost their

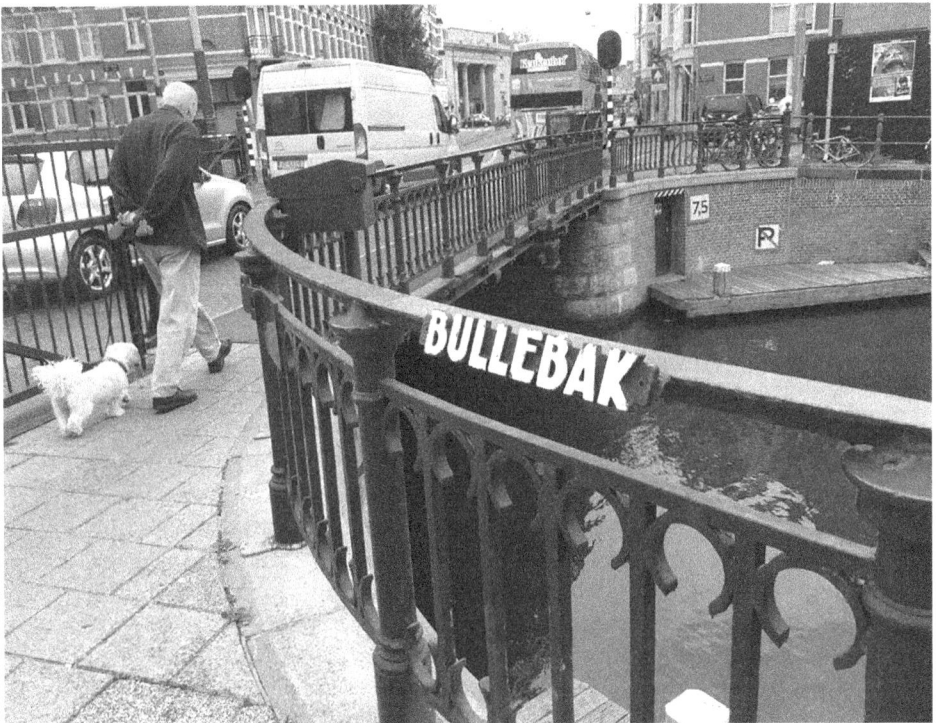

Figure 6.1. The BULLEBAK bridge (photo courtesy of the author)

warehouse function. Years of neglect left them in dilapidated conditions. In the 1970s and '80s, an alternative community of relatively famous and well-to-do artists, dancers, and musicians moved into the neighbour-hood, and numerous decrepit depots metamorphosed into elegant apartments. The reinvented warehouses of the Brouwersgracht inaugu-rated a new architectural industrial aesthetic. The calm of the canal in its midst became an indispensable asset; the dark, smooth water added an intimate atmosphere to the place, connecting both sides, reflecting their lively lights at night. Many of the new folks were gay, and many died – or were dying – of AIDS in the late 1980s. "Taken by the Bullebak," it flashed through my head when I biked over the BULLEBAK bridge.

I moved away to study in the United States. I don't remember the Bullebak ever crossing my mind, even when I spent long summers at home in the Netherlands. As an environmental philosopher, I developed an interest in the social-political and cultural aspects of water. Urban rivers in particular triggered my fascination: they embody a complex dynamic in which any simple – say, clear and distinct – separation between nature and culture dissolves. Furthermore, the reconnecting of cities to their waterways seems to afford a great potential for raising environmental awareness, developing what I have elsewhere called (Klaver 2012a; 2015) an environmental imagination, a place-based social, political, and environmental commitment around public bodies of water, an occasion to learn about and enjoy urban waterways. However, soon it became clear that a city's reconnecting with its water bodies usually comes with a haunting shadow, a spectre, a new kind of Bullebak: rampant, mainly market-driven, waterfront development, this Bullebak does not prey on little kids but on affordable housing.

In the early 1980s, I was part of the Amsterdam squatter movement, occupying houses that were deliberately taken out of circulation for real estate speculation. In those days, housing speculation was an unbridled practice of a handful of wealthy real estate developers, undermining the city's relatively well-organized and fair 1960s housing policy. Amsterdam's twenty-first-century city politics has seen a sharp move towards a neoliberal model, and urban renewal projects around water have contributed to further undercutting affordable housing. Developing waterfront areas has become a lucrative investment, playing a major role in gentrification. Housing prices have skyrocketed, making it virtually impossible for members of the lower and middle classes to own a house or pay the rent. I have written about similar processes in Los Angeles, Dallas, and other places. In the summer of 2016, I began writing an article about Amsterdam and found myself back on the BULLEBAK bridge.

Standing on the BULLEBAK bridge, overlooking the Brouwersgracht, I realized that this canal was one of the first gentrifying places in Amsterdam – even though we did not call it that in those days. The refurbishing of the warehouses along the canal had changed the Brouwersgracht into one of the most expensive and desirable housing areas. It dawned on me that the Bullebak of Amsterdam is not *in* the water; it *is* the water. The water jacks up housing prices considerably. Amsterdam's water monster is alive and well; it is dangerous, and not only for children.

The stories of the dangers of the Bullebak in my mother's cautionary tale provide an anchor point for further considerations of gentrification.

Monster

Of course, water is not a monster in itself. Water is always in a context, a situation. Water creates, affords, washes away, destroys. It generates situations; provides an atmosphere. Thus, the atmosphere around the water-filled ditch of my childhood entailed the Bullebak. A Bullebak arises in, or *is*, the entanglement of materialities, stories, situations, and processes. If there were no children, there would be barely any danger in the ditch, and there would not have been a Bullebak. There would just have been a ditch. A ditch dug for specific functions: to de-water or drain the land and to form a separation barrier, just like a fence or a stone wall. The ditch could be annoying, like when a cousin took the corner too wide, too hastily, and swung into the ditch, bike and all; or after my dad's birthday, when an uncle stumbled out of the doorway and forgot to take a right turn, only to walk straight into the ditch. Wet, muddy, and stinky, folks re-emerged cold and covered with green duck-weed. The ditch was part of the stories that turned into family lore – the laughingstock of birthdays to come. The monster, the Bullebak, was for us, the kids. In relationship to us, the water of the ditch was dangerous; we could drown. Therefore adults told cautionary tales to keep us away from the water – keeping us safe, but also leaving us more drawn to it, intrigued by the ditch.

As grownups, we are supposedly too sophisticated for water monsters. However, in a social-political-economic situation of a housing policy mainly propelled by a private property-driven real estate market, water can become dangerous again, if not monstrous. I am not talking about drinking water, storm water, waste water – all potentially hazardous when mismanaged, and hence crucially important to be subjected to strict regulations. Here, I am talking about the dangers of specific *bodies* of water, such as, rivers, canals, bays, lakes, seas, oceans. They tend to generate highly prized waterfront property. It is fascinating to see ships go by; it is calming to see the sun setting in the sea, to hear the waves of the lake lapping at the shore, to feel the fresh air coming from the water. If we leave housing solely to the workings of market mechanisms, without solid public policy and legislation, living along the water's edge will be only for the affluent few. The water's edge is increasingly turning into private space, or controlled – surveillance space. The Bullebak has morphed into a twenty-first-century water monster: a behemoth of gentrification of waterfronts. The gentrification Bullebak has snuck up on us; he hid in the duckweed of restored buildings, investment banking real estate, chambers of commerce boardrooms.

In *Arts of Living on a Damaged Planet* (2017), Anna Tsing and colleagues assert that monsters are "the wonders of symbiosis and the threats of ecological disruption" (M2). "The monsters in this book," they continue, "have a double meaning: on one hand, they help us pay attention to ancient chimeric entanglements; on the other, they point us to the monstrosities of modern Man. Monsters ask us to consider the wonders and terrors of symbiotic entanglement in the Anthropocene" (M2).

The Bullebak occupies this double meaning: in its man-fish-water constellation it has an ancient chimeric power; as human-created carrier of gentrification around water bodies it reveals "the monstrosities of modern Man," with the threats of ecological disruption and community disruption in its wake. These trends only become more complex in an era of climate change. Threats of sea-level rise and intensification of hurricanes may cause a reverse gentrification trend at waterfronts; now towards a re-development of the higher grounds where the lower-class service providers (used to) live.

In *Hydraulic City: Water and the Infrastructures of Citizenship in Mumbai*, Nikhil Anand (2017) develops a notion of hydraulic citizenship predicated upon the intertwinement of the dynamic infrastructural water flows in pipes and pumps, with citizens, technicians, politicians, and plumbers. He emphasizes the political power of stories: "Stories have multiple vocalities and multiple sites of production. Unlike discourses, stories are particularly attendant to the diverse locations at which human agency is thwarted or dreams are partially realized. Stories are unstable ... The telling of stories is always a political act" (vii–viii).

Following the story of the Bullebak, from the intimate practical-familial to the social-political theoretical, reveals different relations and intertwinements. Mom's monster was an expression of an understanding of the danger of water in the context of children and their thinking and acting in the presence of a ditch. The Bullebak of gentrification exposes the dangers of laissez-faire economic liberalism in development projects around urban waterways. This last Bullebak grabs more than children: it undermines a fair urban riversphere, keeping the less affluent from living and thriving at the water's edge. Gentrification is a situational water monster: it spews unregulated urban planning, unfair housing policy, overpowering Anand's (2017) hydraulic citizen's voice.

Engaging a monster requires understanding what a monster is; how it comes to be, how it spreads its tentacles; what intended and unintended consequences bubble up from it; what the nature of its territory is; and how we understand ourselves in relation to it. As Donna Haraway (1991) states, "Monsters share more than the word's root with the verb 'to demonstrate': monsters signify" (226). Clarifying the workings of the monster reveals the alterity that entails what Haraway (2004) calls the promise of monsters; that is, by clarifying the workings of the monster, the possibility of precaution, resistance, and reframing arises.

The *With*: Riversphere, Meandering, Being in Common

Given their relational nature, a crucial ingredient in engaging with monsters entails an epistemology, ontology, politics, and ethics of the "with." Elsewhere (Klaver 2005) I have analysed the importance of the "with" in these domains, especially for environmental thought. The "with" in ancient Greek is ξυνοσ, better known in its connecting prefix form συν, in Roman alphabet *syn/sun-* as in "synergy"; *sym/sum-* as in "sympathy" or "symbiosis." Travelling through time and various languages, the ξ turns into *s* or *c* and the *u* and *y* mutate into *o*. Thus, the Latinate manifestations are *cum/com-* as in "common" or "community," or *con*, such as "connection" or *co-* like "co-existence" (403). ξυν-, syn-, cum, and *com* all mean "together, with, common, shared." "With" is a small word of great importance, connoting connectivity, complexity, relatedness.

Water provides a medium and framework or theoretical model to think the materiality, sociality, and modes of movement of the "with." I employ two water – specifically river water – dimensions for my "with" framework, namely, riversphere and meandering. To further explicate the workings of the "with," I also weave in the notion of "being in common."

I use the concept of riversphere to expand an understanding of rivers as places of multi-scalar and multi-vector connectivity and complexity. Rivers are more than blue lines on a map. They are more than basins, watersheds, or drainage areas. They influence the geology around them, the air around them, life around them, cultures around them (Klaver 2012b). They create their own hydrospheres, biospheres, and atmospheres. They form intricate networks of relations, conditions of possibilities. My sense of riversphere resonates with Böhme's (1993) concept of atmospheres: "Atmospheres are indeterminate above all as regards their ontological status. We are not sure whether we should attribute them to the objects or environments from which they proceed or to the subjects who experience them. We are also unsure where they are. They seem to fill the space with a certain tone of feeling like a haze" (114). Riversphere also resonates with the notion of ambiance, the cosmopolitan and open ambiance of a city (Amin, Massey, and Thrift 2000) and with Sloterdijk's (2014) notion of globes. The notion of riversphere as atmosphere provides a theoretical frame for thinking the 'with' concretely: it adds social, political, cultural, aesthetic, and emotional dimensions to our thinking about rivers and cities. Riversphere is a thick concept, it includes materialities, concepts, stories and experiences. It enriches rivers in the cultural imagination, spanning the most concrete to the most ephemeral.

The notion of meandering offers a further conceptualization of the workings and movements of the "with." Meandering is based upon the movements of sedimentation and reactivation of water and silt. *Exactly* what happens is unpredictable: many factors and vectors co-determine just what happens. With emergent behaviours and emergent properties of the system as a whole, meandering forms a non-deterministic system: complex, interdependent, and dynamic system, like the weather, or an ecosystem. As such, it symbolizes the wandering motion of exploration.

Meandering derives its name from an actual river, the Meander – now Menderes – River in Anatolia, Turkey. In a meandering of history, the Meander River played major roles in antiquity, and then all but disappeared from the cultural imagination (Klaver 2014; 2016). From early modernity onwards, rivers were straightened – meanders engineered away – in an endeavour to create a more controlled riverine environment. More predictable rivers enabled commercial river traffic operations on precise timetables, exact and relatively stable property delineations, and hence facilitated city planning. In this context, meandering acquired a negative connotation, synonymous with aimless wandering, ambling along a winding path, and rambling through a long-winded argument.

In the course of the second half of the twentieth century, however, one can see an implicit re-evaluation of meandering (Klaver 2014; 2016). Non-linear systems became widely accepted in the sciences; chaos theory and non-deterministic non-linear modelling – meandering – led to valuing complexity in the cultural imagination. Meandering conveys the nature of the non-linear, symbolically and metaphorically; it allows for ambiguity and hybridity, for a rethinking of progress through complexity. Meandering makes room for thinking in terms of atmosphere, for what cannot easily be measured, for the sometimes slow or unpredictable workings of the material realm not ruled by the structures of scheduled time. Meandering bespeaks the social-political necessity of taking time to explore terrains, to elucidate attributes, relations, problems, and solutions, as a gateway to new constructs of imagination, to a capacity to aspire (Appadurai 2004).

Theories of complexity are well-suited to a twenty-first-century era of high technology, globalization, and urbanization: "With its many convergent, overlapping and irreversible interdependencies 'globalization' is remaking 'societies' but not in a linear, closed and finalized form. We might see the growth and spreading of theories of complexity as part of, and simultaneously helping to enact, the very processes of global change" (Law and Urry 2005, 404). Within a riversphere and meander approach, simple geometrical models of nature and city planning are expanded with models of indeterminacy and complexity (Klaver 2017), by models of flows; not just flows of water, but of people, capital, light, luggage, tourists, money, exchanges, and experiences.

As a third element to unpack the workings of the "with" in an urbanizing globalizing world, I invoke David Harvey's (2008) emphasis on the importance of the *common* in his conceptualization of "right to the city":

> The question of what kind of city we want cannot be divorced from that of what kind of social ties, relationship to nature, lifestyles, technologies and aesthetic values we desire. The right to the city is far more than the individual liberty to access urban resources: it is a right to change ourselves by changing the city. It is, moreover, a common rather than an individual right since this transformation inevitably depends upon the exercise of a collective power to reshape the processes of urbanization. (23)

The crucial word here is "common." The centre of Harvey's right to the city is the shift from individual to the "with," the common.

Together, riversphere, meandering, and the conceptualization of the city as a place of being in common, lead me to a three-fold model of the 'with.' I employ this model to engage Amsterdam's water monster, the Bullebak of gentrification, and facilitate an understanding of its character, thus gesturing to what Donna Haraway calls, the *promise* of monsters (Haraway 2004) – that is, the possibility of resisting. In this case, this entails addressing the injustices of gentrification.

In the following section, I bring these concepts into play through an historical description of water developments in Amsterdam.

Amsterdam: Dam in a Watery Place

One thousand years ago, there was nothing that even hinted at a beginning of Amsterdam. There was soggy, waterlogged land – spongy peat riddled with streams. The first people who settled the area were able to stay only through labour-intensive and collaborative processes of reclamation. They faced permanent adjustment to the water: an ongoing battle forcing them to come up with new ways to cooperate and to construct ditches and dikes – always bigger and higher, stronger and longer. At a certain point, they had too much invested in the land to walk away from it (Feddes 2012, 27–8). Out of this muddle of water, land, and stubborn ingenuity, Amsterdam arose.

A millennium later, the initial water hardship has been turned into an advantage: the city's 2015 report *Watervisie [Water Vision] Amsterdam 2040* opens with "Amsterdam is a water town. Its original name was Amestelledamme, medieval Dutch for 'Dam in a watery area or place'" (my translation) (3). In the foreword, Alderman Kock (for Water) and Alderman van der Burg (for Spatial Planning) emphasize that water is Amsterdam's "identity," its "allure." They proclaim proudly, "*Het water in Amsterdam is voor iedereen!*" [*Water in Amsterdam is for everyone!*] (my translation). The italicized message also decorates the front page of the report (see Figure 6.2).

Amsterdam's water definitely *was* for everyone. A thousand years ago, everyone had to deal with it, live with it, work it, if not fight it. However, in the twenty-first century, it is a big question – maybe *the* question – whether Amsterdam's water, in its manifestation of public space, still *is* for *everyone*, i.e., still is a being in common. In a predominantly neo-liberal political climate, urban space around water has become a valu-able asset, prone to gentrification. A market-driven economy is more conducive to a city for the few than for the many (Amin, Massey, and Thrift 2000). In *The Just City* (2010), Susan Fainstein determines three conditions for urban justice "within the wealthy Western world": "equity, democracy, and diversity" (165). The question "Is water in Amsterdam for everyone?" becomes a pressing question on precisely these three lev-els. Who is able to use the water and the water's banks (how equitable is the distribution and use?), how democratically are decisions being made, and how diverse is the population that can actually enjoy living by Amsterdam's water?

Amsterdam is often heralded for its consensus-building policy, its well-coordinated public investments, and "the strength and continuity of its urban-planning capacity" (Healy 2007, 37), culminating in a much acclaimed "model of considerable justice" (Fainstein 2010, 140). This might have rung true for a large part of the twentieth century, but the characterization has been jeopardized – partly because of the success of Amsterdam's water. A relatively fair housing *policy* has increasingly been replaced with a hard-core real estate *market.*

Early in the twenty-first century, Amsterdam joined the bandwagon of cities marketing themselves to stay competitive as a major city for business and tourism. A 2003 report, *Choose Amsterdam*, found that the "Dimensions of Business city, Knowledge city and Residence city" needed to be "Strengthened" (3), and in 2004, the logo "I *amster-dam*" was launched (*The Making of the City* 2004, 11), echoing New York City's successful slogan "I love New York." Even though it does not seem to be a recipe for success to replace loving with being, the new branding caught on. The logo has become a city icon; the sleek large-scale letters in white and red at highly travelled places such as the Rijksmuseum and Schiphol Airport are popular photo hotspots (see Figure 6.3).

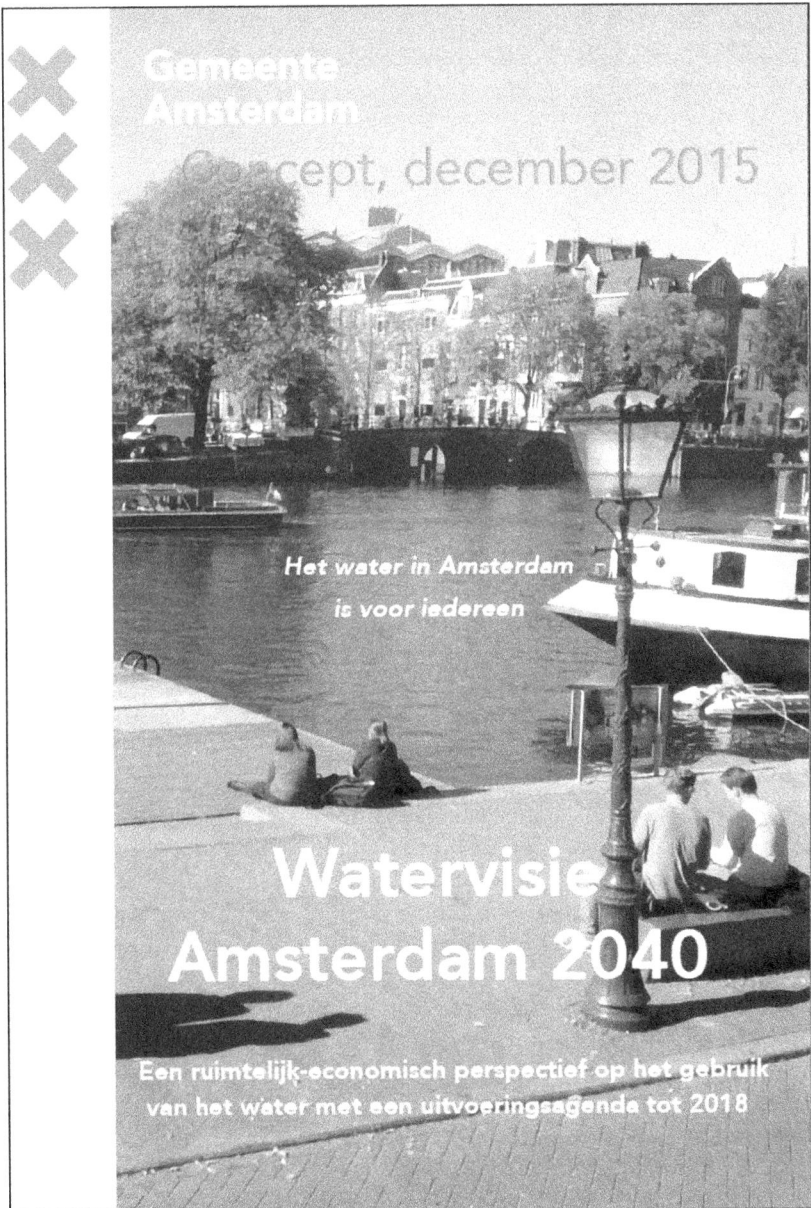

Figure 6.2. Cover of the city's 2015 report: *Watervisie Amsterdam 2040* (photo courtesy of the author)

Figure 6.3. I amsterdam in front of the Rijksmuseum (photo credit: Kim De Wolff. For more images, see the official site: www.iamsterdam.com/en/)

The public–private partnership in the city branding took off: by mid-2016, Amsterdam had attracted so many tourists that it decided to temporarily suspend marketing. Regarding the city's tourism policy, Alderman Ollongren (for the economy) stated in early 2016 that "the goal is to slow down the growth" (Couzy 2016).

The twenty-first century is the century of the city. In 2007, the global urban population, for the first time in history, surpassed the rural population. According to the 2014 United Nations report *World Urbanization Prospects,* 66 per cent of the world's population is projected to be urban by 2050. European urban population will be over 80 per cent in 2050 (it was 73 per cent in 2014). The report emphasizes that sustainable development challenges will concentrate in cities and will require integrated policies (UN DESA 2014, 1–7).

These trends require us to rethink and reinvent urbanism, in which cities can be agents of change rather than mere engines of growth, change for greater social justice and environmental sustainability. How we imagine cities in this new century is of critical importance; so, too, how we envision "urban citizenship" (Amin et al. 2000) and, in our context, a citizenship in relation to water, something Anand (2017) called "hydraulic citizenship."

Cities form a practical and theoretical base for building new connections and cross-connections. Participatory models of global democracy, with cities as crucial nodes, could, according to Barber (2013), make politics more nimble than the nation-state: "Let cities, the most networked and interconnected of our political associations, defined above all by collaborations and pragmatism, by creativity and multiculture, do what states cannot" (4). Recognizing the fundamental importance of rivers to cities yields a fertile formula for a new urbanism, forging further partnerships and collaborations.

Strictly speaking, Amsterdam is not a river town; experientially and functionally, it is. It has created itself as a river town, and is constantly recreating itself as such. A significant part of its economy, urban planning, architecture, aesthetics, and tourism is built around a riverine existence. The trio of concepts – riversphere, meandering, and the being in common – provides a critical lens on river cities. It opens up questions such as who and what is benefitting from cities reconnecting to the rivers, and who and what is suffering (Kibel 2007)? Who or what processes are making the decisions, who is gaining what, and who is sacrificing what? Let us further explore how Amsterdam became an urban node in a larger water world.

Hybrid Waters from Aa to IJ

A, the first letter of the alphabet, is also Frisian for a body of water, Aa/Ee. Water pervades Dutch history. The Netherlands is the result of an age-long intertwinement of agential processes of land, water, and people; all living and working *with* each other. Amsterdam's soil consists of "alternating layers of clay, sand and peat, permeated with fresh and salty groundwater" (Feddes 2012, 63). A rocky substratum is to be found only 400 metres deep. Where many European cities have underground "shadow cities," with caves, catacombs, and tunnels hewn out of the substratum, "Amsterdam floats" (63).

Amsterdam's "river," the Amstel, was a natural-artificial construct. It was not much of a river – no comparison to other European rivers, such as the Rhine, Danube, Seine, Thames, Tiber, or Volga. According to Feddes (2012):

> It is even an open question whether you could or can call the Amstel an independent river. Many streams meandered through the delta landscape, running into one another and sometimes changing their course, and they cannot always be isolated as separate rivers. To this day "Amstel" is still the designation for a rather randomly bounded part of this network of small streams. (10)

The Amstel's name is derived from *Aeme-stelle*, old Dutch for "water-place" (*Aeme-*, "water"; *stelle*, "area" or "place").

The first settlers appeared around the year 1000 CE. Between 1200 and 1250, they connected two small peat streams with a canal, reversing the direction of the southern stream northwards, discharging into a larger water body, the IJ. This newly connected watercourse became the river *Aemstel*. As Feddes (2012) argues, "The Amstel we know today is therefore not a natural river but a hybrid construction, half natural and half creation of man, and it is precisely this artificiality that gives it its historical power" (28).

This artificiality, this hybridity, is a recurring theme of Dutch water in general and Amsterdam in particular.

Around 1275 CE, the two banks of the Amstel close to its mouth at the IJ were connected with a dam, and consolidated the village *Aemstelredamme*.

Amsterdam quickly grew larger; its waters occasioned a flourishing shipbuilding and seafaring culture. The IJ (see Figures 6.4 and 6.5) provided easy access to the Zuiderzee and from there to the North Sea. By the seventeenth century, thousands of Dutch commercial sailing ships were traversing the world's oceans – their well-organized international trading networks spanned the globe. At the southern tip of Chile, they sailed from the Atlantic into the Pacific Ocean, naming this crucial point at the Tierra del Fuego archipelago Cape Horn after the West-Frisian Zuiderzee town Hoorn. Were the West-Frisian Bullebak stories, that meant to frighten generations of kids to stay away from the rural ditches, part of the cargo that made its way to Amsterdam? At least two sluices in Amsterdam of the seventeenth and eighteenth centuries were popularly called Bullebak; they were dangerous places with irregular and intense water flow. The West-Frisian Bullebak now clearly had taken up residence also in Amsterdam.

The Netherlands became the leading seafaring nation, a significant colonial power, and entered its Golden Age. Amsterdam, with its bustling port, became the wealthiest city in the world, the beating heart of a newly emerging capitalist economy, home to the world's first stock exchange. Its population boomed and affluent merchants invested capital in one of the largest urban expansions of the time, the *Grachtengordel*, or Canal Belt. As David Harvey would have predicted, surplus capital was invested in real estate and land development.[1]

The *Grachtengordel*, the iconic semi-concentric ring of parallel canals around the city's centre, was an ingenious water-based construction, another example of the Dutch working and living *with* water. It was the signature of the "free burghers," the new class of prosperous traders, snubbing the aristocracy: the Herengracht (the Canal of the "Heren" the higher echelons of the burghers) far surpassed the other canals architecturally, leaving the Keizersgracht (Emperor's Canal) and Prinsengracht

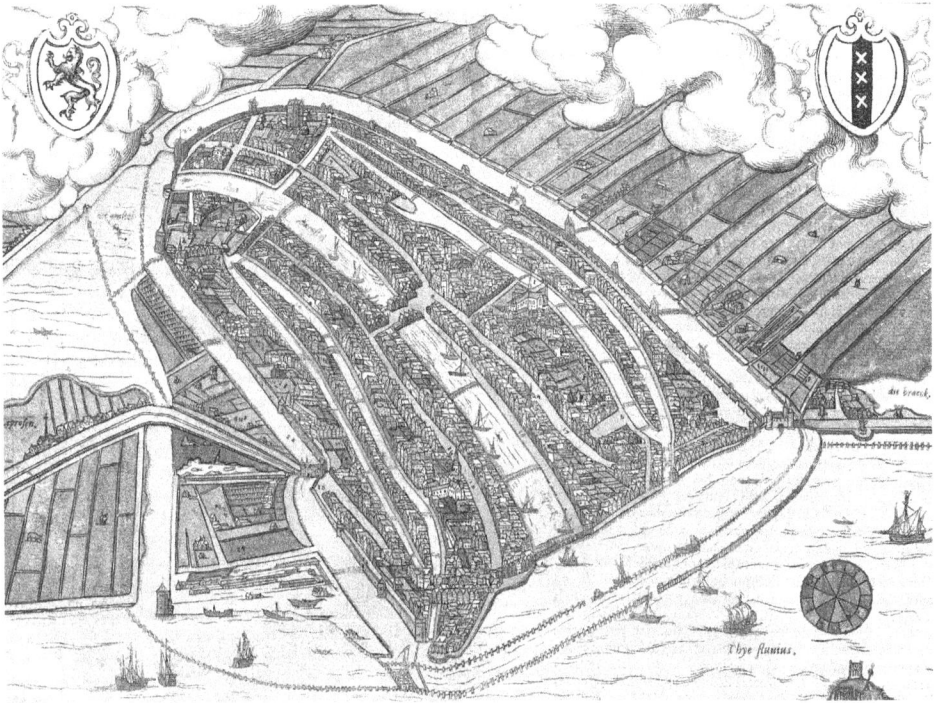

Figure 6.4. Map of Amsterdam in 1567, with the dam in the centre and sailing ships in *Thye Fluvius* (Latin for "the IJ river") (University of Amsterdam Library)

(Prince's Canal) inferior, explicitly symbolizing the economic superiority of the burgher over the aristocracy. The translation of Herengracht in tourist guides as Lord's Canal or Patrician's Canal misses the point of the snubbing. Characteristic gables and façades crown the houses displaying the new wealth. Most of them are now historical monuments. As Harvey argues, right from the beginning there was a solid relation between capitalism and urbanism (see Figure 6.5).

In 2010, the *Grachtengordel* was placed on the UNESCO World Heritage List. Artificiality was again a crucial motif, explicitly mentioned in the UNESCO category of Outstanding Universal Value: "The Amsterdam Canal District is the design ... of a new and entirely artificial 'port city.' It is a masterpiece of hydraulic engineering, town planning, and a rational programme of construction and bourgeois architecture. It is a unique and innovative, large-scale but homogeneous urban ensemble" (UNESCO 2010, criterion (i)).

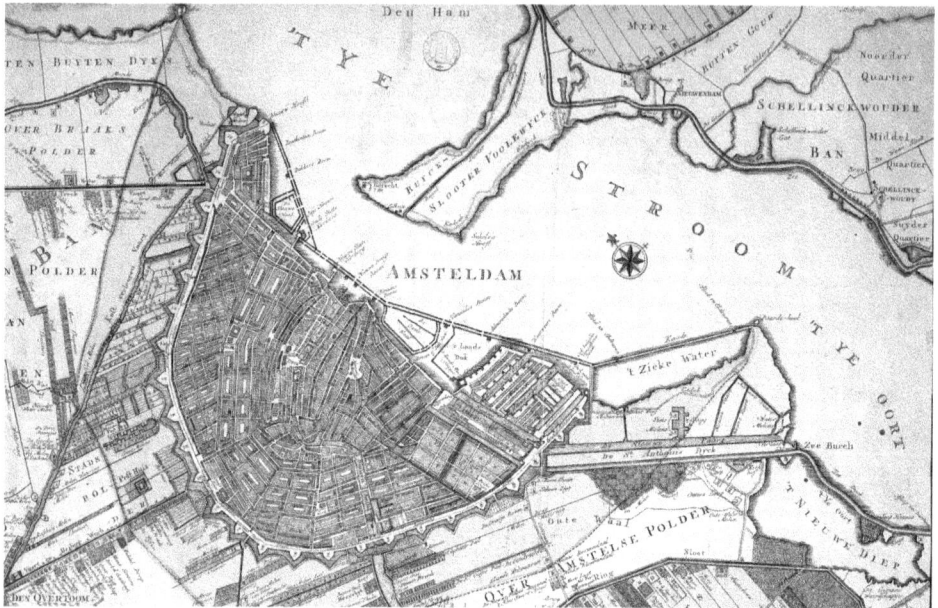

Figure 6.5. Map of *Amsteldam* (Amsterdam) 1770 with *'t Ye Stroom*, the IJ stream (University of Amsterdam Library)

Not the "river" Amstel, or the IJ, but the *Grachtengordel* with its picturesque atmosphere has put Amsterdam in the cultural imagination and on tourist maps as the "Venice of the North." According to *Watervisie Amsterdam 2040* (2015), canal cruises are one of the most successful tourist attractions of the Netherlands, drawing three million visitors a year (6).

Water seems to be the background, the stage, but it is in fact the materiality that made the rise of trade possible in the seventeenth century. Sailing ships could carry goods over oceans, traversing great distances, and barges transported heavy loads over rivers and canals, thanks to the ways of water, its buoyancy, its natural properties, its behaviour.

Like the emergence of the Amstel as a river, the resurgence of the IJ as a major water body was partially "artificial" – in this case, an *unintended* consequence of human activity, human interventions afforded by the water. In that sense, we can speak of a *co*-creation: humans and water working *with* each other, shaping the place *together*. It is unclear exactly what the IJ looked like between 800 and 1000 CE, just before the first settlers arrived in the area that is now Amsterdam. There has been no conclusive research about whether in those days the IJ was an open water body, relatively closed peat land, or a partially open waterway (de Bont 2008, 360–70). What is known with certainty is that early settlers had to dig extensive drainage ditches into the peat to make the soil dry enough for agriculture. Over the years, the cumulative effect of these small-scale reclamations across the Dutch delta led to serious soil subsidence. Much of the land ended up beneath sea level; when massive storms hit North Holland in the second half of the twelfth century, it flooded severely. This so-called natural disaster was deeply rooted in human hydraulics. One of the consequences was the reappearance of the IJ. After the first ice age, the IJ had emerged as a major sea arm, but became filled in with vegetation over the ensuing centuries. After the floods, only a layer of dunes stopped the IJ from returning to its marine origins.

On early maps, it was called *Y* or *Ye*, a now obsolete word for water, rooted in the West-Frisian word for "stream; small river," derived from Germanic **ahwō*, "water," cognate with the Frisian *Ae/Ee*, names – as I noted above – for bodies of water, as in *Aemestelle* or *Amestelledam*, or the French word for water, *eau*.

The IJ is sometimes referred to as a lake, but this categorization is disputed. It is a hybrid water body: part lake, part river, part human construct. *Rijkswaterstaat*, the water management and public works branch of the Ministry of Infrastructure and the Environment, officially calls it a river, a designation we saw on the older Amsterdam maps above: the 1567 map used the Latin name for river, *fluvius: Thye Fluvius*, and the 1770 map mentioned *'t Ye Stroom*, the IJ stream. As the waterfront of Amsterdam, the IJ definitely is *experienced* as a river.

In 1876, the IJ was connected again to the North Sea, this time through the North Sea Canal, a straight and reliable connection for the shipping industry. The sand from digging the canal through the dunes was deposited into the IJ at the mouth of the Amstel to create islands on which to build Amsterdam's Central Station. When the train station was ready in 1889, the inner city's direct access to the IJ was gone. A centuries-old relation had ended.

This meant that the Port of Amsterdam needed to be relocated: it was moved westwards, on the banks of the IJ and the North Sea Canal. In 1951, the Amsterdam-Rhine Canal was created to connect the IJ to the crucial shipping artery of the Rhine River. The canal is now one of the busiest shipping canals in the world, and Amsterdam port a major transport hub.

Amsterdam Meandering Back to the IJ

The Central Station was built, controversially, with its back to the IJ, cutting the city off from its major waterway. During most of the twentieth century, the area between the Central Station and the southern bank of the

Figure 6.6. Swimming in the IJ, July 2015 (photo courtesy of the author)

IJ was a motley mix of warehouses, storage companies, and railroad emplacements interlaced with marginal strips of deserted space, pockets of lost land left for city nomads, squatters, artists, and the dark life of heroin and prostitution. Most industry and shipyards had moved to the northern bank of the IJ and moved away altogether in the latter part of the twentieth century, when the rise of aviation and global shifts in shipbuilding radically changed the character of the industry (de Hoog 2012, 41).

For centuries, the IJ had mainly an industrial and infrastructural role: the gateway to the world of trade. At the end of the twentieth century, it became a key player in urban development. With its broad waterway and extensive barge traffic, it has the allure of a major river, but is close enough to the city centre to piggyback on an intimate atmosphere of urban living. With its water quality safely regulated, it even became possible to take a dip in the IJ (see Figure 6.6). Real estate has been booming at the banks of the IJ; in another meandering, or twist of history, some call Amsterdam Noord, the northern IJ bank, "the Manhattan of Amsterdam."

The banks of the IJ are developing into high-end places, with a mixture of public and private buildings, such as the EYE Film Museum, the Music Building, Passenger Terminal Amsterdam, the public library, and

Figure 6.7. Cranes at the IJ Boulevard, 2010 (photo courtesy of the author)

the IJ boulevard (see Figure 6.7). Now, in the twenty-first century, even a new view of the position of the Central Station is emerging, where the station opens up to the renewed IJ (Schram, van Ruyven, and van der Made 2012).

Major manifestations are organized around the IJ: Its most famous is SAIL Amsterdam, established in 1975 to celebrate the city's 700th anniversary. Every five years, for five days in August, SAIL attracts millions of visitors. It has become one of the largest maritime festivals in the world: the biggest sailing ships from around the world participate – hundreds of majestic historical ships, welcomed by 10,000 smaller boats (see Figure 6.8).

Living with the Bullebak

Amsterdam is meandering into a new future, following the drum of a booming real estate market. Urban reconnecting to waterways usually comes with gentrification and a complex redrawing of the public and the private: when the old, neglected, polluted, dangerous riverside becomes "beautified" and safe, the older, usually rather poor neighbourhoods are elbowed out to make room for a new upper middle-class population

Figure 6.8. Flagship SAIL Amsterdam 2010, the *Stad Amsterdam* (photo courtesy of the author)

(Klaver and Frith 2014). This is not to say that abandoned, polluted, and dangerous places should not be changed into liveable spaces. The question, however, is how democratically this is organized. For Harvey (2008), "The freedom to make and remake our cities and ourselves is ... one of the most precious yet most neglected of our human rights" (23). He emphasized that this is a common right rather than an individual right: a collective power to reshape the processes of urbanization. As Kibel (2007) states succinctly: With any of these projects we need to "consider the questions of who makes decisions about our urban rivers ... and who ultimately benefits from or is burdened by these decisions" (15).

Besides gentrification, the danger of commodification is a sanitized and controlled space – lacking the conceptual and social "messiness" not only of abandoned and waste places, but also of economically more diverse communities. The real estate beat is occasionally punctuated with local and political resistance. Rivers are political; they have long been places of contestation: the word "river" is etymologically related to the word "rival." How are modern-day conflicts and demands defined? In terms of ecology and diversity? Quest and conquest of the riverbanks entails searching together *with* each other, reclaiming the meaning of *con* in conquest. No longer can conquest be a neo-colonial occupation and privatization; rather, it is a public endeavour, fighting for a city council that supports fair housing policy. This challenges homogeneity, commodification, privatization of public space; it means an active policy for multicultural opening of the riverbanks to create a heritage of inclusion, a complex vibrant riversphere.

Riversphere employs the dynamic power of meandering to enable collective and agile thinking and planning, fostering an atmosphere to commit to the vision of the *Watervisie Amsterdam 2040* report: *Water is for everyone!*

Amsterdam, with a long tradition of collaboration, needs to revitalize its engagement with water through deliberative models, models of democratic decision, many of them rooted in a long history of living and working *with* water. Only thus can the city ensure that the monster of market mechanisms does not become too powerful and drag many of us away. That requires an ethics, politics, and philosophy of being and thinking *with* – with everyone involved: citizens, water, water creatures, water monsters, the Bullebak.[2]

NOTES

1 In "Authentic Landscapes at Large: Dutch Globalization and Environmental Imagination" (Klaver 2012a), I show how Amsterdam merchants spent a considerable amount of their surplus capital in land reclamation projects, the Dutch polders, which created a radical change in the Dutch landscape experience. I argue that it led to the rise of the famous so-called Dutch school of realism in landscape painting, rooted in seascapes. I show that in fact the "realistic" painters painted with their backs to the changing landscape, focusing on the old rivers and dunes.

2 Acknowledgements: Thanks to Brian C. O'Connor for his many suggestions, close reading, and editing; Kim De Wolff for editing suggestions; René Boomkens for literature recommendations; and Ton Dekker for always bringing me home to the meanderings of Amsterdam.

REFERENCES

Amin, Ash, Doreen Massey, and Nigel Thrift. 2000. *Cities for the Many Not the Few.* Bristol, UK: Policy Press.

Anand, Nikhil. 2017. *Hydraulic City: Water and the Infrastructures of Citizenship in Mumbai.* Durham, NC: Duke University Press.

Appadurai, Arjun. 2004. "The Capacity to Aspire: Culture and the Terms of Recognition." In *Culture and Public Action*, edited by Vijayendra Rao and Michael Walton, 59–84. Stanford: Stanford University Press.

Barber, Benjamin. 2013. *If Mayors Ruled the World: Dysfunctional Nations and Rising Cities*. New Haven: Yale University Press.

Böhme, Gernot. 1993. "Atmosphere as the Fundamental Concept of a New Aesthetics." *Thesis Eleven* 36 no. 1: 113–26. https://doi.org/10.1177/072551369303600107.

De Bont, Chris. 2008. *Vergeten land. Ontginning, bewoning en waterbeheer in de westnederlandse veengebieden (800–1350)*. Dissertation, Wageningen University. Accessed 02 March 2018. https://www.wur.nl/en/Publication-details.htm?publicationId=publication-way-333639363034 and http://library.wur.nl/WebQuery/hydrotheek/1889255/.

Couzy, Michiel. 2016. "Amsterdam gaat toerisme beteugelen door strengere toetsing nieuwe hotels." Het Parool, 9 January. Accessed 10 August 2016. http://www.parool.nl/parool/nl/5/POLITIEK/article/detail/4221093/2016/01/09/Amsterdam-gaat-toerisme-beteugelen-door-strengere-toetsing-nieuwe-hotels.dhtml/.

Fainstein, Susan S. 2010. *The Just City*. Ithaca, NY: Cornell University Press.

Feddes, Fred. 2012. *A Millennium of Amsterdam: Spatial History of a Marvelous City*. Bussum, The Netherlands: THOTH.

Haraway, Donna J. 1991. "The Biopolitics of Postmodern Bodies: Constitutions of Self in Immune System Discourse." In *Simians, Cyborgs, and Women: The Reinvention of Nature*, Donna Haraway, 203–30. New York: Routledge.

– 2004. "The Promises of Monsters: A Regenerative Politics for Inappropriate/d Others." In *The Haraway Reader*, 63–124. New York: Routledge.

Harvey, David. 2008. "The Right to the City." *New Left Review* II no. 53: 23–40.

Healey, Patsy. 2007. *Urban Complexity and Spatial Strategies: Towards a Relational Planning for Our Times*. New York: Routledge.

De Hoog, Maurits. 2012. "Amsterdam en het IJ." In *Amsterdam terug aan het IJ, transformatie van de Zuidelijke IJ-oever*, edited by Anne Schram, Kees van Ruyven, and Hans van der Made, 35–41. Nijmegen: Uitgeverij SUN.

Kibel, Paul Stanton. 2007. "Bankside Urban: An Introduction." In *Rivertown: Rethinking Urban Rivers*, edited by Paul Kibel, 1–22. Cambridge: The MIT Press.

Klaver, Irene J. 2005. "The Implicit Practice of Environmental Philosophy." In *Environmental Philosophy: Critical Concepts in the Environment*, Vol. IV: *Issues and Applications*, edited by J. Baird Callicott and Clare Palmer, 398–408. New York: Routledge.

– 2012a. "Authentic Landscapes at Large: Dutch Globalization and Environmental Imagination." *SubStance: A Review of Theory and Literary Criticism*, 41 no. 1: 92–108. https://doi.org/10.1353/sub.2012.0007.

– 2012b. "Placing Water and Culture." In *Water, Cultural Diversity & Global Environmental Change: Emerging Trends, Sustainable Futures?* Edited by Barbara

Rose Johnston, Lisa Hiwasaki, Irene J. Klaver, A. Ramos Castillo, and Veronica Strang, 9–29. UNESCO International Hydrological Program. Dordrecht: Springer Press.

– 2014. "Meander(ing) Multiplicity." In *Water Scarcity, Security and Democracy: A Mediterranean Mosaic*, edited by Francesca de Châtel, Gail Holst-Warhaft, and Tammo Steenhuis, 36–46. Ithaca, NY: Cornell University Press.

– 2015. "Accidental Wildness on a Detention Pond." *Antennae: The Journal of Nature in Visual Culture* 33 (Autumn): 45–58.

– 2016. "Re-Rivering Environmental Imagination: Meander Movement and Merleau-Ponty." In *Nature and Experience: Phenomenological Approaches to the Environment*, edited by Bryan E. Bannon, 113–29. London: Rowman and Littlefield.

– 2017. "Indeterminacy in Place: Rivers as Bridge and Meandering as Metaphor." In *Phenomenology and Place*, edited by Janet Donohoe. London: Rowman and Littlefield.

Klaver, Irene J., and J. Aaron Frith. 2014. "A History of Los Angeles's Water Supply: Towards Reimagining the Los Angeles River." In *A History of Water, Series III, Vol. 1: Water and Urbanization*, edited by Terje Tvedt and Terje Oestigaard, 520–49. London: I.B. Tauris.

Law, John, and John Urry. 2005. "Enacting the Social." *Economy and Society* 33 no. 3: 390–410. https://doi.org/10.1080/0308514042000225716.

The Making of the City: Marketing of Amsterdam. 2004. Amsterdam: City of Amsterdam.

Schram, Anne, Kees van Ruyven, and Hans van der Made. 2012. *Amsterdam terug aan het IJ, transformatie van de Zuidelijke IJ-oever.* Nijmegen: Uitgeverij SUN.

Sloterdijk, Peter. 2014. *Globes: Spheres II.* Los Angeles: Semiotext(e).

Tsing, Anna, Heather Swanson, Elaine Gan, and Nils Bubandt, eds. 2017. *Arts of Living on a Damaged Planet.* Minneapolis: University of Minnesota Press.

United Nations, Department of Economic and Social Affairs, Population Division. 2014. *World Urbanization Prospects: The 2014 Revision, Highlights (ST/ESA/SER.A/352).* New York: United Nations. Accessed 10 August 2016. https://esa.un.org/unpd/wup/Publications/Files/WUP2014-Highlights.pdf/.

United Nations Educational, Scientific and Cultural Organization. World Heritage Centre. 2010.

UNESCO. "World Heritage Committee Inscribes Five New Cultural Sites on World Heritage List and Approves Two Extensions to Existing Properties." Accessed 10 July 2018. http://whc.unesco.org/en/news/643/.

– "Seventeenth-Century Canal Ring Area of Amsterdam Inside the Singelgracht – UNESCO World Heritage Centre." Accessed 10 July 2018. http://whc.unesco.org/en/list/1349.

Watervisie Amsterdam 2040 (Concept). 2015. Roy Berents and Hans Straver. Amsterdam: Ruimte en Duurzaamheid (R&D) of the City of Amsterdam.

7 Water and the City: Towards an Ethos of Fluid Urbanism

INGRID LEMAN STEFANOVIC

For years, architects and urban planners have known that using water elements in city environments creates a better quality of urban life.

– Wallace J. Nichols

More than just supporting an improved quality of life, water holds the promise of seeing and engaging the city in profoundly new ways. While on some level a consequence of "a planning of pipes, of roadworks and accounting," the city is also so much more (Lefebvre 2007, 138). In fact, as an instantiation of our very being, it can be said that the city is also the penultimate incarnation of human dwelling and a place where "values come to their most concrete expression" (Grange 1999, xv).

Cities are conceived in a multitude of ways – as market centres; as transportation hubs; as architectural artifacts. As human artifacts, they are often distinguished from the wildness of nature. But cities are no more simply physical *things* than they are some-*thing* that exists independently of the natural world. More and more, the nature/urban divide is seen to collapse in new, integrative visions of *natural cities* (Stefanovic and Scharper 2012). No matter how hard we try, nature has its way of quietly infiltrating cities on many different levels (Manià 2017; Van der Ryn and Cowan 1996).

This chapter suggests that cities, understood as more than an assemblage of static elements, are themselves shaped by the volatility of water. Seen from the perspective of fluidity, the promise of a new ethos of urbanism emerges through a sensibility to the interplay of notions of dynamism, relationality, spatio-temporal reverberations, evocative memories, aspirations, transformations, and a concomitant search for serenity and moments of calm.

Part One describes how the experience of water opens the possibility of seeing cities differently. As Marcel Proust reminds us, "the real voyage of discovery consists not so much in seeking new territory but possibly in having new sets of eyes" (cited in Nichols 2014, 269). Seeing water as more than a calculative resource invites reflection on the meaning of the city as an integral element of both the human and natural worlds.

Part Two considers the practice of placemaking. How does the architecture of water inform water and architecture? What does the relation between water and architecture tell us about how to build in a way that incorporates ethical, aesthetical, and ontological moments in praxis? Two narratives of place – one that captures water's fluidity and motion, the other its stillness – aim to explore how water is integral to sense of place.

Part Three raises the prospect of a new ethos of water – one that aims to bring the conversation to the realities of policy making and decision-making practices. Part Four summarizes some of these reflections and looks ahead at how we might imagine future prospects of city building in light of water's presence.

I: The Natural City: Thinking with Water

That cities are no longer to be defined independently of environmental pressures is increasingly a given among the broader public. Nevertheless, while we more commonly talk about "sustainable," "green," and "smart" cities, a dualistic paradigm, deeply rooted within the Western metaphysical tradition, continues to exert an influence as we slip into distinguishing between the human versus the ecological, the economy versus the environment, and cities versus nature.[1]

Water invites us to rethink such oppositions and to focus instead on the essential belonging of these apparently contrary notions. Without water, neither human beings nor ecological systems can be sustained. Economies have prospered precisely because they are sustained in built environments adjacent to water bodies. And cities themselves, no less than the natural world, grow and develop thanks to the grace of water sources that infuse them.

Attending to water moves us beyond dualistic scenarios, sensitizing us to the realities of natural flux, *physis*, and inviting us to reflect beyond the reductionist parameters of calculative thinking (Heidegger 1966). Water reminds us to think outside a narrowly anthropocentric worldview and to contemplate our place within a watery world whose ontological givenness we did not create. Just reflecting upon the fact that we ingest and excrete water daily – the same element that has passed through geological time – helps us to envision human existence as less one of mastery but rather one of belonging and being beholden to nature itself.

Often taken for granted and forgotten in our daily undertakings, water at the same time remains so very close, sustaining us as living beings. Infiltrating our environments and constituting 55 to 60 percent of the adult body, water is existentially definitive of who we are.[2]

Beginning our reflections by pondering metaphors of water – metaphors "rich to the point of paradox" – may help us to see the infusion of the natural in the urban through different eyes (Moore 1994, 16). Water is *no one thing*: it is life-giving – a symbol of fertility, chastity, youth – but also an evil force of death in that it "relentlessly dissolves bonds, it spoils, it drowns, it wears away, it rots, and it floods. It even unravels memory" (17). In that sense, it reminds us that our own lives are fragile, ultimately themselves vulnerable, finite, and in flux.

On some level, water is a composition of diverse moments: "There is often too much water and there is often too little water. Water purifies, water needs to be purified ... Water disorients and reorients" (Ashraf 2014, 92). Symbolizing "neither terra nor firma," it invites us to contemplate phenomena of "Immersion. Buoyancy. Drift. Level. Depth. Fluidity. Flotation. Ebb. Tide. Rhythm" (ibid.).

Definitively capturing a single interpretation of water in itself is a fluid, elusive process. However, water can certainly be represented in reified, scientific terms as a chemical molecular formula – one oxygen and two hydrogen atoms. It can be framed as a utilitarian resource, technically controlled and managed through engineered infrastructure and quantified along numerous parameters. Such descriptions, legitimate on so many levels, are examples of a modern, technological, calculative paradigm that informs most policy making in the Western world.

Still, water itself invites a different way of thinking – what Heidegger called "meditative" or "originative" thinking (Heidegger 1966; Stefanovic 2000). Philosophers have acknowledged how, before constructing abstractions (e.g., water = H_2O), we are in the world in a more primordial way through the immediacy of lived experience and embodiment. Our encounter with water itself opens us up to a way of seeing the world that makes that evident.

Writers often note that "water touching our skin is the most personally intimate experience we can have of it" (Moore 1994, 202). Kong Jian Yu describes how his visceral childhood knowledge of water as "interconnecting" and "about continuities" was disrupted for him in a science lesson that taught that water was colourless, tasteless, and shapeless – a vision that left him and his classmates confused:

> How can water be colorless? The White Sand Creek to the west of our village is white, while the Black Dragon Lake in the forest south of our village

is blue. How can water be tasteless? We drink water from the spring by the creek, and it is sweet ... How can water be shapeless? Our language and literature has taught us that the Chinese character of "water" has the shape of a winding river escorted by ribbon-like lakes. And what about the colorful pebbles, dancing grass and swimming fish in the water? Don't they have anything to do with water?

Too often, abstract scientific or utilitarian descriptions of water remain on the level of calculative thinking, ignoring the full depth of the lived experience of water. To discover the meaning of water, theologian Matthew Fox invites us to go without it for three days. "The first sip of water then becomes a very sacred act."[3] That wonder in experiencing water as life-giving exceeds calculation or quantifiable value, reminding us of our finitude and beholdenness to the natural world for our very existence.

Scientists themselves are recognizing that a spontaneous, existential draw to water is deeply rooted, neurologically, to ways of creative thinking that are informed through non-linear, fluid patterns. Moving in multiple directions, waves and ripples permeate our embodied ways of being differently than hard, rough lines that lead to analytic understanding, psychological inflexibility, and, according to some, even a decline in creative thinking (Nichols 2014, 202). Flowing water is often seen to reflect prethematically the streaming of time itself. Confucius saw water as an unending stream, "like what passes," time itself (Allan 1997, 36). Phenomenologist Gaston Bachelard similarly felt that "in his inmost recesses, the human being shares the destiny of flowing water ... A being dedicated to water is a being in flux" (Bachelard 1983, 6).

It is this fundamental, ontological draw to water that helps to explain the results of a 2010 study by Plymouth University researchers in the United Kingdom. Subjects were asked to rate over 100 pictures of natural and urban environments and gave higher preference ratings to virtually *any* picture containing water (Nichols 2014, 12). Globally, water views are said to impart a trillion-dollar premium on real estate, while reputable estimates suggest that 80 per cent of the world's population lives within 100 kilometres of an ocean, lake, or river coastline (Nichols 2014, 9, 70). Even the colour blue is reportedly favoured three or four times more over other colours by people around the world (Nichols 2014, 87).

That we are drawn, on a visceral level, to water is reported frequently in the literature. That draw is more than simply the consequence of a rational, instrumental decision to locate near water to facilitate shipping between markets or to support basic physiological functions. Water infiltrates our environment, but also, on a deep ontological level, our embodied minds. In that sense, water is not simply an external element of "nature," independent of who we are as humans. Nor is it independent of how we dwell in our cities, which are themselves embedded in a world of nature and water.

II: Designing with Water in Mind

In his classic text on "Building Dwelling Thinking," Heidegger (1971) reminds us that the design and construction of our settlements is more than a simple matter of calculative fabrication or deliberate, functional production. Rather than a heavy-handed process of assembly, building as dwelling invites an attitude of sparing. "Real sparing," we learn, "takes place when we leave something beforehand in its own nature, when we return it specifically to its being, when we 'free' it in the real sense of the word into a preserve of peace" (149).

Heidegger uses the example of building a bridge to show how it provides more than a simple utilitarian function of connecting two riverbanks that serve as "indifferent border strips of the dry land" (1971, 152). Instead, the bridge "designedly" causes the banks of the river "to lie across from each other" (152). He explains how construction of the bridge "brings stream and bank and land into each other's neighborhood. The bridge gathers the earth as landscape around the stream. Thus, it guides and attends the stream through the meadows" (152).

On the one hand, from a calculative perspective, we can see a bridge as an engineered artifact in and of itself. Indeed, structurally, it must stand on its own functional integrity. On the other hand, much more can be said about such functional integrity, recognizing that any individual construct is part of a larger context, properly reflecting the interplay of diverse elements and spatio-temporal moments. In the words of Henri Bortoft, "a part is a place in which the whole can be present" (1996, 12). In that sense, a building or bridge is never simply an independent, isolated object. It is, inevitably, always part of a larger context within which it finds its appropriate fit.

Arising within this sensibility is an approach to building that already invites a markedly different approach to creativity and design. Rather than being a Faustian, manipulative production of discrete entities, the process of city building suggests replacing willful construction with a new capacity to see, listen, and discerningly respond to more subtle messages within broader contexts.

Paying heed to water's presencing within our environments and our embodied selves may provide an opportunity to further re-envision the design and city-making process itself. Becoming aware of the place of water may open the possibility of entirely inverting our design thinking to "reread architectural and urban conditions through nature's material continuities and environmental complexities, rather than the reverse" (Berman 2014, 113). Such an inversion entails a growing sensibility to natural cities that themselves incorporate an urbanism that is more fluid as well as adaptive: rather than relying simply upon a reductionist strategy of engineered storm drains, we can incorporate natural drainage through innovative rain garden designs or draw upon wetlands in place of lawns and sewage treatment plants, thereby working with the natural movement of water rather than aiming to dominate it.

The German poet and thinker Johann Wolfgang von Goethe understood how water's way of being was instructional. According to Goethean phenomenologists, "water speaks a language of movement," adopting variable forms with "a remarkable degree of order as if it had a life and intention of its own" (Seamon and Zajonc 1998, 233, 235). Attending to water's natural rhythms opens up the possibility of designing through "flowforms" – channels designed to enable such rhythmical movements, revealing water in its dynamism and aiding in water purification, transport, irrigation, and aquaculture.

Such presencing of water invites us to attend to architectural building and city planning as moments of what we might envision as fluid urbanism: designing with water in mind – with a sensibility and responsiveness to the dynamic, elemental, natural sustenance provided by water as it essentially infiltrates our lives and the built places wherein those lives find meaning. In the words of Marion Weiss, "to leverage [water's] truest value, we must design more multivalent, supple, resilient strategies to collect, cleanse, distribute, cultivate and ultimately enjoy the fundamental nature and luxury of water ... and maybe, we must design with water as if our lives depended on it" (2014, 131).

The fact is that water's inviolate attributes serve as a reminder that we dwell within nature's own logic and constraints. Designing with water in mind means being attuned to a "fitting placement" and discerning stewardship of our built and natural environments (Mugerauer 1994, 132ff). Charles Moore is the author of one of the few books dedicated to water and architecture. He reminds us that "the key to understanding the water of architecture is to understand the *architecture of water* – what physical laws govern its behavior, how the liquid acts and reacts with our senses, and, most of all, how its symbolism relates to us as human beings" (1994, 15).

Designing with water in mind sensitizes us to building in a way that invites an appreciation of the grace and complexity of the natural world itself. We do not create water but must uncover a fitting relation to it within our settlement designs. Discerning such a relation summons an attitude of respect and beholdenness in lieu of tyrannical manufacturing; stewardship of broader contextual landscapes instead of the fabrication of singular artifacts; flexibility and humility rather than imposing universalizing, essentializing solutions. In the words of architects who have built waterscapes around the world, "the essence of water cannot be grasped by means of abstract physical and chemical formulas; this conceptualization is simply too static. Water is dynamic and demands a similar mental fluidity in understanding it" (Grau and Dreiseitl 2009, 12).

In fact, what we learn when we attend to water is that successful architectural design and city building are less the product of willful human construction, and more a matter of seeing and listening to the givenness of taken-for-granted natural contexts. "There is an art to looking at water," reflects Sarah Allan (1997, 23). In that sense, "design for water should not be seen through conveyors, containers and conduits but rather as a means to humanize and elucidate its sensation in ways that are performative and prophetic" (Hood 2014, 83). Design for water should be a means to uncovering a more balanced and sensible relation between building and the natural world in which it is situated.

III: Moments in Design: Flowing to Still

How might we illustrate such performative and prophetic moments in city design? Let us consider instances of city planning that have emerged with water in mind.

While there are multiple examples to which one might point, I offer two instances that may be instructive. The first deals with moving waters and the interplay between physical absence – and community presence – of hidden streams. Despite city builders' attempts to subdue and control the visibility and power of water in cities, it remains relentless in its dynamic presence, moving beneath the apparent stability of our dwelling places.

The second example is one of explicitly designing in respect of still water. Modernist architects have, at times, acknowledged that water plays an integral role in building design, helping to define the larger spatial contexts. "Contemplative waters play an important role in mythology" – and in thoughtful urban design as well (Moore 1994, 124).

By way of context, it is important to acknowledge that architects and planners incorporate water into their designs in many diverse ways: "when the fusion of architecture and water is treated carefully and creatively, the potential for meaningful expression is practically limitless" (Moore 1994, 22). Not only visually but aurally, the presence of water in our dwelling places introduces, on a deeply visceral level, a moment of connectivity to the natural world to which we are biologically, aesthetically, morally, and often pre-thematically drawn.

Architects have learned how, much as with visual water elements, its sounds define dwelling places. Frank Lloyd Wright's renowned "Fallingwater" – today a national historic landmark – near Pittsburgh, Pennsylvania, was commissioned by Edgar J. Kaufmann in 1937 and built to stand suspended above a 9-metre waterfall. The client expected the house to be built along the riverbanks, with the waterfall serving as backdrop. Wright, however, felt that it was more important to integrate the home more closely with the waterfall so that it would serve as an integral, structural element of its design. In his words, it was key "not simply to look at the waterfalls but to live with them" (1988, 43).

And so, significantly, the waterfall's presence is not actually seen from the home's interior, but heard: it is the sound of the rushing water that defines its closeness (Moore 1994, 202). In Wright's words,

Fallingwater is a great blessing – one of the great blessings to be experienced here on earth. I think nothing yet ever equaled the coordination, sympathetic expression of the great principle of repose where forest and stream and rock and all of the elements of structure are combined so quietly that really you listen not to any noise whatsoever although the music of the stream is there. But you listen to Fallingwater the way you listen to the quiet of the country. (1988, 43)

Other houses have been constructed not above water, but directly upon it. Ijburg – the Netherlands' first floating neighbourhood – hosts 97 floating homes. Concrete foundations are filled with styrofoam, supporting houses that are deemed to be "unsinkable." According to Koen Olthuis, an architect who designs such structures, "sustainability on the water can work even better than sustainability on the land" (Shamsion and Pineda 2015).

In fact, "amphibious architecture" is proving to be a sustainable solution for some communities that dwell in floodplains. While designed to normally rest on the ground, amphibious foundations allow for homes to be temporarily elevated, in response to rising floodwaters in areas that do not suffer high currents or waves (English 2014, 21). From New Orleans to the Netherlands, amphibious homes are already proving the technology, and plans are in place to develop similar housing projects in flood-prone regions of Nicaragua and Bangladesh, as well as in the United States and Canada (22). These homes – no less than other fascinating examples, such as floating parks and floodable public squares – provide noteworthy examples of architecture that work as a fitting response to the natural world, operating in a way that is discerning and mindful of water's ways (Taylor-Foster 2014; Scott 2017).

Incorporating water into architectural designs can help to block undesirable noise, recharge our minds, reduce urban stress, and sustainably support innovative community design. Indeed, waterscapes can be and have been meaningfully integrated into urban plazas, parks, city master plans, residential projects, river restoration projects, interior designs, fountains, and civic art projects (Grau and Dreiseitl 2009).

A particularly interesting example of integrating moving waterscapes into urban areas is the daylighting of streams. Historically often converted to underground sewer canals, these waterways are being rediscovered in cities around the world, sometimes – as in the case of the Cheonnggyecheon Stream Restoration Project in Seoul, South Korea – resulting in increases of corridor biodiversity of over 600 per cent (Grau and Dreseitl 2009, 114ff; Landscape Architecture Foundation 2012).

But streams need not be physically daylighted to bring water close to the community. A case in point is Taddle Creek, in downtown Toronto, Canada. Originally a sacred Ojibway gathering place, the full 6 kilometres of the creek were eventually buried in 1884 to address growing wastewater and sewage contamination issues.

Physically burying the stream was not, however, the end of the story. Over the decades, communities have been inspired to revive the legacy of the creek. Daylighting plans for the stream, which flows under the University of Toronto campus grounds, have been percolating for decades, stymied due to lack of funding. Aiming to preserve the waterway's legacy, a Montessori school carries the name of Taddle Creek; a community magazine bears its name; and Taddle Creek Park in the Annex neighbourhood hosts a sculpture in honour of the stream that remains buried along its entire length.

In some respects, one of the most visible testimonials to the lost river is a central piece of artwork in Taddle Creek Park. *The Vessel* is a water fountain, a 5.7-metre-high (18.5-foot-high) sculpture in the form of a giant jug (Figure 7.1). Constructed from 4 kilometres of stainless steel tubing, it is meant to symbolize the distance of the stream from the park to its final destination in Lake Ontario. According to some, that tubing also serves as a reminder of the Indigenous people who travelled the creek by canoe, although neither the distance represented through the steel structure nor the creek's Indigenous roots are evident (Stefanovic 2013a).

One feature – a cascading waterfall over the lip of the jug – does disclose water as moving, as streaming. The park's architect, Paul Martel,

Figure 7.1. *The Vessel* (photo courtesy of Oriana Leman)

describes how "the water flows quietly from the ring of jets at the top of the Vessel, down the sides of the body, and falls on the paving laid in concentric rings at its base, where it pools and is captured in drains and recycled to flow down the Vessel again. Children love to splash in the ponded water." Particularly at night, Martel believes, the lighting "at certain times gives 'The Vessel' an otherworld appearance of translucence, as it seems to levitate above the paving." Even its stainless steel tubing is, according to Martel, a "non-solid" reminder of a lost river – "part myth but also a real watercourse somewhere under the park" (Stefanovic 2013a).

While *The Vessel* is intended to artistically represent a lost river, one might argue that the stream has shown itself in more meaningful ways on its own. In the early 1930s, the underground flow caused the foundations of a major hotel to unexpectedly shift; engineers had to attend to the building, which eventually leaned a full 15 centimetres towards a major road artery. Similarly, in 1985, Metro Toronto Police headquarters' construction project was met with a surge of water from the creek's tributary, necessitating a redesign of the building. Community planner Eduardo Sousa points out how "the Taddle makes itself known on the Bloor subway lines too; many sump pumps and other engineered solutions have been installed to deal with the Taddle there," just as flooded basements in homes during heavy rains bring the stream to light (Stefanovic 2013b). One homeowner describes how water, silt, and leaves bubble up into his basement in a storm: "My household river has become a fixture in my consciousness, an ever-present reminder that there is an upstream and a downstream" (Chernos 2010). As much as we try to keep the waters submerged, they rise to the surface unexpectedly.

Landscape architect Michael Hough's work is an interesting example of a planner acknowledging and working together with the buried stream. Years ago, Hough led a Lost River walk focused on Taddle Creek. The Lost River walks, a joint initiative of the Toronto Green Community and the Toronto Field Naturalists, were founded by philosopher Helen Mills as a way to reconnect local communities with their forgotten watersheds. Since 1995, more than 17,000 people have joined over 560 walks, rediscovering hidden urban waterways.

According to Mills, Hough used the opportunity to explain his redesign of a portion of the University of Toronto campus called "Philosopher's Walk," situated directly above Taddle Creek. While he had originally hoped to dig up and daylight this section of the creek, he was unable to – it had been converted into a functioning combined sewer. So instead, in his role as landscape architect, he "unfilled" the landscape and recreated the original shape of the ravine. "It used to be flat," Mills explains, "but he purposely re-landscaped it to show the presence of the ravine and the creek that used to flow there" (Stefanovic 2013c).

While the water is physically hidden, the phenomenological experience of its presence is lived meaningfully – and often pre-thematically – through the slope and shape of the landscape. It is memorialized through plaques along walkways. The community continues to bear witness to the creek's legacy in its naming of newspapers, schools, and parks. And sculptures and testimonials explicitly seek to ensure that the water is honoured through the creativity of artifacts.

A second, very different example illustrates water's presence as a moment of stillness in built spaces. It shows how the explicit incorporation of water into architectural design is integral to sense of place. One cannot imagine certain spaces in the absence of water: the meaning and identity of such built forms are specifically defined with water in mind.

One such case arises on the campus of Simon Fraser University, in British Columbia, Canada. The Burnaby "Academic Quadrangle" is a Canadian landmark, designed by internationally renowned architects Arthur Erickson and Geoffrey Massey. A quintessential example of modernist architecture, the core of the campus was built in 1965, anchored within a central pond (see Figure 7.2). Erickson – described by a former governor general of Canada as "the greatest architect we have ever produced" – is said to have found inspiration in the Acropolis and the hill towns of Italy, where the mountain on which the university sits – and its surrounding landscape – were explicitly incorporated into the design, and terraced buildings follow the contours of the terrain (Martin 2009). In the words of Phyllis Lambert, founder of the Canadian Centre for Architecture, Arthur Erickson did no less in his design of the SFU campus than create "architecture of the earth, out of the earth," with the mountain literally dictating the shape of the campus (Martin 2009).

Much has been written about the terraced, horizontal building designs and even the structure of the hallway-like quadrangle, explicitly aimed at encouraging interdisciplinary collaboration. Little attention, however, has been directed to the central pool that defines the space in a core manner. Inattention to that central water space seems a strange oversight in a province defined by its rainforest climate, on a campus constructed to ensure that one remains dry and undercover as one navigates among the buildings.

Charles Moore reminds us that "in its capacity to reflect, the artificial pool can be a compelling compositional device" (1994, 127). "[S]tationary waters are natural reflectors: their mirrored surfaces absorb, repel and refract their surroundings. Mirrored images of landscapes ... expand space by ... repeating the infinite depth of blue skies" (124).

Figure 7.2. Simon Fraser University, Burnaby Campus (photo courtesy of the author)

And so, too, here at Simon Fraser University's central pond, the draw is to a depth and sheen that reflects both the surrounding architecture as well as the clouds and the sun. Pools of water such as these "clean out an unoccupiable space in front of our buildings so that we can view them free of more mundane components" (Moore 1994, 201). Moore has pointed out that "if the reflecting pool at the Taj Mahal were drained and planted with grass, the tomb would lose a great deal of its mystery" (1994, 201). The main campus of Simon Fraser University would suffer no less a loss in the absence of this central artificial pond. The water is a place for contemplation, quiet, and reverie, reminding us that despite the concrete presence of human surrounds, the natural world is above and below and around us. Large orange fish weave their way below the surface. A massive rock face interjects itself above the still waters. Seated beside the pool, one is invited to daydream in silence. Each such moment serves as a testimonial and a reminder that while people, artifacts, and knowledge define the central university campus, so does the natural presencing of water interweave itself into our lifeworld in a vital, organic manner.

Perhaps, in that sense, our built environments can only be genuinely meaningful moments of design to the extent that they draw us towards a remembrance of our beholdenness to the natural world itself. Creativity in building is less a matter of projecting innovative styles than discerning appropriate relations. Finding our proper place in cities means, therefore, recovering our relation to earth, as well as to water. It means "generating form and guiding design with the environment as an active participant in the process" (Gattegno 2014, 31).

IV: Looking Ahead – Towards an Ethos of Fluid Urbanism

Water holds significance on many different levels, from the biophysical to the ontological, aesthetic, and spiritual. It also invites reflection on moral grounds – on how one might build the good life in respect of environmental parameters and the presence of water. In that regard, philosophers have begun to develop a field of "water ethics" – encompassing all manner of traditional theories from utilitarianism to deontology in the search for moral guidance in decision making about water issues (Brown and Schmidt 2010; Llamas, Martinez-Cortina, and Mukherji 2009; Groenfeldt 2013).

The issues raised in this chapter invite further reflection on two levels. First, how does a sensibility towards water, as a moment of lived experience, inform that moral conversation? A second, not unrelated question relates to how urban design is to be integrated into the discussion of water ethics.

As a start, let us acknowledge the integral connection between ethics and building. The fact is that how we build reflects and informs our systems of values and our worldviews. Artifacts themselves have been said to "have politics" inasmuch as "important ideas and central values may be embedded and expressed in the built environment" (van den Hoven 2013, 136). Increasingly, there is a call to attend to "value-sensitive design": "incorporating moral values into the design of technical artifacts and systems by looking at design from an ethical perspective" (137). In that respect, some argue that, already, there has been a "design turn in applied ethics," meaning that

> design is a respectable ethical category. Instead of taking human character of a person's actions as the unit of analysis and the object of moral evaluation, it seems sometimes highly relevant to be able to ask questions about the moral quality of a design. We need to be able to evaluate proposals to change the world and undertake this evaluation from the point of view of moral values. (van den Hoven 2013, 138)

In that respect, the normative function of architecture is to express cultural values, as they are embedded in our built spaces, as well as to constrain certain types of behaviour through mindful design practices (van den Hoven 2013, 136–7; Stefanovic 2000). Attending to the moral dimensions of place might reveal how spaces protect or, alternatively, jeopardize privacy and safety; how they might reflect authoritarian political ideals; how they might encourage a sense of belonging or, on the other hand, a sense of disorientation (Newman 1972; Stefanovic 2000; Stefanovic and Scharper 2012). Our architecture is never morally neutral: arguably, the effect on our sense of who we are and what cultural ideals we support are very much embedded in and affected by the community structures and places that we build.

In that case, how might water inform the conversation about such a design turn in applied ethics? Certainly, seeing water differently does not automatically generate a set of moral rules or principles to be followed in some form of mathematical precision. Rather than speaking about water *ethics,* perhaps a different discussion emerges from understanding moral judgment as more than a set of abstract, rational deductions. Given that we are embodied and ontologically engaged in the world, perhaps such engagement opens up the possibility of developing a new water *ethos* – a sensibility towards the centrality of water in all of our ways of being and dwelling.

According to some, "it is imperative that a new design discourse must consider a 'water ethos,' where water is the measure and instrument of spatial and social organization. A water ethos indicates a fundamental immersion in water with anthropological and sociological significance" (Ashraf 2014, 92). Approaching architectural and place design within the framework of such a water ethos means being open to resilience, change, fluidity, possibility, dynamism, and complexity, rather than restricting ourselves to static visions of reified structures. It means bringing an awareness of water and natural elements into a view of organic building that "*develops* from within, outward in harmony with the conditions of its being, as distinguished from one that is *applied* from without" (Wright 1988, 7). After all, "places which persist in rigidity (be it in planning mentality or civic perception) do not correspond to the character of water. Water spaces and places are capable of shaping themselves and have the capacity of perpetual self-renewal" (Grau and Dreiseitl 2009, 12).

Incorporating such a vision of fluid urbanism offers the promise of creating communities that advance moral sensibilities of the good life in the deepest ways. In the end, such an ethos of fluid urbanism holds the promise of advancing an innovative and genuinely thoughtful vision of natural cities (Stefanovic and Scharper 2012).

Hopefully, such a vision and sensibility towards water in urban design will have contributed something, in the words of architect Frank Lloyd Wright, "to make life where we were and as we lived it better, brighter and more beautiful" (1988, 200).

NOTES

1 Witness how we may seek to acknowledge these dualisms: on the environment/economy debate, see https://www.theglobeandmail.com/opinion/editorials/on-economy-vs-environment-team-trudeau-is-finding-the-balance/article32166280/ and on the nature/city divide and why "getting back to nature" matters, see http://thespiritscience.net/2016/08/02/nature-vs-cities-how-nature-literally-restructures-your-brain-patterns-2/. Accessed 25 February 2018.

2 See https://water.usgs.gov/edu/propertyyou.html/. Accessed 25 February 2018.

3 Reported in an interview on "Insights on the Edge": http://www.soundstrue.com/podcast/transcripts/matthew-fox.php?camefromhome=camefromhome/.

REFERENCES

Allan, Sarah. 1997. *The Way of Water and Sprouts of Virtue*. Albany, NY: State University of New York Press.

Ashraf, Kazi. 2014. "Water as Ground." In *Design in the Terrain of Water*, edited byAnuradha Mathur and Dilip da Cunha. Pennsylvania: Applied Research + Publishing, with the University of Pennsylvania School of Design), 91–7.

Bachelard, Gaston. 1983. *Water and Dreams: An Essay on the Imagination of Matter*. Dallas, TX: Dallas Institute Publications.

Basta, Claudia, and Stefano Moroni, eds. 2013. *Ethics, Design and Planning of the Built Environment*. New York, London: Springer.

Berman, Ila. 2014. "Inundation to Scarcity." In *Design in the Terrain of Water*, edited by Anuradha Mathur and Dilip da Cunha, Pennsylvania: Applied Research + Publishing, with the University of Pennsylvania School of Design, 113–21.

Bortoft, Henri. 1996. *The Wholeness of Nature*. New York: Lindisfarne Press.

Brown, Peter G., and Jeremy J. Schmidt, eds. 2010. *Water Ethics: Foundational Readings for Students and Professionals*. Washington: Island Press.

Chernos, Saul. 2010. "Creek in My Sewer Pipe." *Now* magazine, 7 October, 30: 6. Accessed 20 August 2017. http://www.nowtoronto.com/news/story .cfm?content=177103/.

English, Elizabeth. 2014. "Amphibious Architecture: An Innovative Strategy for Flood Resilient Housing," *Canadian Risks and Hazards Network*, 6 no. 1 (Fall), 20–2. Accessed 25 July 2017. http://docplayer.net/45416733-Haznet -volume-6-no-1-fall-2014.html/

Gattegno, Nataly. 2014. "Aqueous Territories." In *Design in the Terrain of Water*, edited by Anuradha Mathur and Dilip da Cunha.Pennsylvania: Applied Research + Publishing, with the University of Pennsylvania School of Design, 31–9.

Grange, Joseph. 1999. *The City: An Urban Cosmology*. Albany, NY: State University of New York Press.

Grau, Dieter, and Herbert Dreiseitl. 2009. *Recent Waterscapes: Planning, Building and Designing with Water*. Boston: Birkhauser Verlag AG.

Groenfeldt, David. 2013. *Water Ethics: A Values Approach to Solving the Water Crisis*. New York: Routledge.

Heidegger, Martin. 1966. *Discourse in Thinking*. New York: Harper & Row.

– 1971. "Building Dwelling Thinking." In *Poetry, Language, Thought*. New York: Harper & Row, 143–61.

Hood, Walter. 2014. "Common Sensing." In *Design in the Terrain of Water*, edited by Anuradha Mathur and Dilip da Cunha. Pennsylvania: Applied Research + Publishing, with the University of Pennsylvania School of Design, 83–9.

Landscape Architecture Foundation. 2012. "Cheonggyecheon Stream Restoration Project." Accessed 20 September 2017. http://lafoundation.org /research/landscape-performance-series/case-studies/case-study/382/.

Lefebvre, Henri. 2007. "Philosophy of the City and Planning Ideology." In *Philosophy and the City: Classic to Contemporary Writings*, edited by Sharon M. Meagher. Albany, NY: State University of New York Press, 136–9.

Llamas, M. Ramon, Luis Martinez-Cortina, and Aditi Mukherji, eds. 2009. *Water Ethics*. London: Taylor & Francis.

Manià, Kirby. 2017. "'A Garden Had Been Left to Grow Wild There': Considering Nature in Ivan Vladislavić's Johannesburg." *Safundi* 18 no.1, 69–84. https://doi.org/10.1080/17533171.2016.1252144.

Martin, Sandra. 2009. "The Greatest Architect We Have Ever Produced." *The Globe and Mail*, 21 May. Accessed 15 October 2017. https://www.theglobeandmail.com/news/national/the-greatest-architect-we-have-ever-produced/article4275827/?page=all/.

Moore, Charles W. 1994. *Water and Architecture*. New York: Harry N. Abrams.

Mugerauer, Robert. 1994. *Interpretations on Behalf of Place: Environmental Displacements and Alternative Responses*. Albany, NY: State University of New York Press.

Newman, Oscar. 1972. *Defensible Space*. New York: Macmillan.

Nichols, Wallace J. 2014. *Blue Mind*. New York: Little, Brown.

Scott, Ellen. 2017. "London Now Has a Super Chilled Floating Park." *Metro News*, 1 June. Accessed 15 August 2017. http://metro.co.uk/2017/06/01/london-now-has-a-super-chilled-floating-park-6676935/.

Seamon, David, and Arthur Zajonc, ed. 1998. *Goethe's Way of Science: A Phenomenology of Nature*. Albany, NY: State University of New York Press.

Shamsion, Jacob, and Chelsea Pineda. 2015. "The Netherlands Is Building Entire Neighborhoods that Float on Water." *Business Insider*, 17 December. Accessed 30 June 2017. http://www.businessinsider.com/netherlands-floating-houses-2015-12/.

Stefanovic, Ingrid Leman. 2000. *Safeguarding Our Common Future: Rethinking Sustainable Development*. Albany, NY: State University of New York Press.

– 2013a. "The Death and Life of a Hidden Stream, Part 3: An Artist's Rendition of Taddle Creek." *Water Canada*, 8 July. Accessed 31 August 2017. http://watercanada.net/2013/part-three-the-death-and-life-of-a-hidden-stream/.

– 2013b. "The Death and Life of a Hidden Stream, Part 4: Taddle Creek and the Violence of the Waters." *Water Canada*, 20 August. Accessed 31 August 2017. http://watercanada.net/2013/part-four-the-death-and-life-of-a-hidden-stream/.

– 2013c. "The Death and Life of a Hidden Stream, Part 5: Taddle Creek and Sense of Place." *Water Canada*, 16 October. Accessed 31 August 2017. http://watercanada.net/2013/part-five-the-death-and-life-of-a-hidden-stream/.

Stefanovic, Ingrid Leman, and Stephen Scharper, eds. 2012. *The Natural City: Re-Envisioning the Built Environment*. Toronto: University of Toronto Press.

Taylor-Foster, James. 2014. "Architecture and Water: Exploring Radical Ideas to Unlock the Potential of Urban Waterways." 23 October. Accessed

4 September 2017. http://www.archdaily.com/559891/architecture-and
-water-exploring-radical-ideas-to-unlock-the-potential-of-urban-waterways/.

van den Hoven, Jeroen. 2013. "Architecture and Value-Sensitive Design." In
Ethics, Design and Planning of the Built Environment, edited by Claudia Basta
and Stefano Moroni. New York: Springer, 135–41.

Van der Ryn, Sim, and Stuart Cowan. 1996. *Ecological Design*. Washington, DC:
Island Press.

Weiss, Marion. 2014. "Cultural Watermarks." In *Design in the Terrain of Water*,
edited by Anuradha Mathur and Dilip da Cunha. Pennsylvania: Applied
Research + Publishing, with the University of Pennsylvania School of Design,
131–9.

Wright, Frank Lloyd. 1988. *In the Realm of Idea*. Carbondale and Edwardsville:
Southern Illinois University Press.

8 What We're Talking about When We're Talking about Water: Race, Imperial Politics, and Ruination in Flint, Michigan

SARAH J. KING

In Flint, Michigan, in 2014–15, the drinking water supply of the entire city was poisoned with lead when the city switched its drinking water source from the City of Detroit to the Flint River without following proper treatment protocols.[1] Almost immediately after the source of Flint water was changed, local residents started complaining about the water quality, including taste, smell, and turbidity. At public meetings, mothers challenged local water managers to defend the quality of the water; the mothers were uniformly dismissed as reactionary and un-educated. For months, people in charge of running the water system (the city) and overseeing it (the state and the Environmental Protection Agency [EPA]) insisted that nothing was wrong. Meanwhile, children, the sick, and the elderly across the city became sicker and sicker, including developing rashes, behavioural problems, hair loss, etc.

Eventually, it was confirmed that the water in the Flint River had not been treated with corrosion control before being pumped through the municipal system, and as it was slightly corrosive, it had begun to wear away at the old lead pipes that were part of the city water system. Not only did the city Water and Sewerage Department not appropriately treat the water, they did not properly test the water, and did not believe local residents when they complained about water quality and illness. Employees of the state and the federal EPA, both responsible for oversight of local drinking water supplies, were complicit in these failures and in related illegal cover-ups. As a result, hundreds of children, and sick and elderly people were poisoned with lead in the City of Flint, and the city entered a public health emergency. The entire city water supply was polluted and toxic; in some places, the lead in the water was high enough to classify the water as toxic waste. National Guard troops were called in and, with Red Cross volunteers and others, delivered bottled water and tap filters door to door to all Flint residents. By summer 2017, fifteen officials in charge of the city water supply, including emergency managers and bureaucrats at the City, the state Department of Health and Human Services, and the Department of Environmental Quality had been charged with crimes, including involuntary manslaughter, obstruction of justice, tampering with evidence, conspiracy, and misconduct in office; the governor's office remains under investigation (Egan 2017). The water supply has since been switched back to the Detroit River source, but the pipes are now corroded and many are ruined, and can no longer safely supply drinking water. The city, working with the state and the federal EPA, poisoned its own residents, destroying not only their trust but also any remaining value in their homes and neighbourhoods.

It's tempting to characterize the ruination of the drinking water system in Flint as a case of "accidental pollution" by people who "made mistakes." After many months of activism on the part of local residents, the first official acknowledgements of problems in the water system framed them in these terms. As the crisis wore on, it began to be seen as a failure of legislation and oversight, and was positioned as such by politicians and bureaucrats. The question became where to hang that failure – on the governor? The state water protection bureaucrats? The EPA? Eventually, the state attorney general also laid criminal charges against former municipal, state, and EPA employees who neglected their responsibilities, falsified reports, and covered up the toxicity of the water, increasing the harm to residents. Victoria Morckel, a geographer at University of Michigan–Flint, portrays the situation as an urban planning failure (2017). Butler, Scammell, and Benson (2016) suggest that we understand it as an example of regulatory failure and environmental injustice. Explanations such as these continue to see the situation in Flint as a failure of law, to some degree – as a situation in which the law did not do what it was intended to do, and in which people charged with upholding the law did not carry out their responsibilities.

Placing the racialized ruination of the Flint drinking water system in the context of the city's own history, and of North American imperialism, demonstrates how the pollution of the drinking water system occurred within the context of a larger system of imperial control designed to maintain the ruination of an already marginal racialized city to protect the well-being of privileged others elsewhere. It is tempting to talk about events such as those in Flint as though they are disconnected from other environmental crises in North America. But failing to see patterns, treating each of these crises as unique failures, continues to place all of the responsibility for these situations on local players, and allows the larger structures of power that rely on and perpetuate the persistent recurrence of such situations to remain invisible.

To illuminate some of the hidden social and political patterns at play, the Flint case is explored in the context of another seemingly dissimilar water conflict that I have discussed extensively in its own terms elsewhere: the fishing dispute at Esgenoôpetitj/Burnt Church, New Brunswick. In 1999, a prolonged and violent fishing dispute erupted at Esgenoôpetitj/ Burnt Church among an Indigenous community, the Mi'kmaq, their settler neighbours, and the Canadian government after the Supreme Court of Canada's *Marshall* decision upheld the treaty right of the Mi'kmaq to fish and earn a moderate living from their catch (King 2014). This dispute got plenty of media attention over many years, and the government and many commentators continued to insist that the dispute was about fish. For local residents, both Indigenous and settler, the conflict was actually grounded in place, and in their contested (post)colonial relationships to their own lands and waters.

Place is a way of understanding the web of interrelationships between humans and the other-than-human world that shapes both humans and the other-than-human through time. Humans are affected by the plants, animals, rocks, mountains, lakes, rivers, or seas where they live, and the presence of humans affects the plants, animals, rocks, mountains, lakes, rivers, and seas. Consider, for example, that human bodies are made mostly of water. We drink and excrete water, eat creatures from water, and use water in all manner of ways in our daily lives. How we go about doing that has a huge effect not only on us, but on water, too – on individual bodies of water, on watersheds, and on the entire water cycle. The value of taking a holistic, place-based approach is to see the ways in which water is connected not only to other elements of the ecosystem, but also to other elements of the social and political systems of humans. This chapter is an attempt to sketch out some of those connections and to raise some of the complex political and justice-oriented questions that arise once we take the dialogic challenge of place seriously.

My book *Fishing in Contested Waters: Place and Community in the Dispute at Burnt Church/Esgenoôpetitj* explores the depth and complexity of the local Indigenous and settler experiences of place and the dispute; I am drawing on that work here to show how race, ruination, and imperial politics are important at Esgenoôpetitj/Burnt Church and in the larger (post)colonial North American context, such as in the ruination of the drinking water system in Flint. I draw this comparison not because the communities of Esgenoôpetitj and Flint are necessarily alike – while the remote Mi'kmaw community and the post-industrial city are both highly impoverished, unemployed, racialized, and segregated, little in the public framing of the crises in these two places is similar. Esgenoôpetitj had a "fishing dispute," a conflict over local employment, Indigenous rights, and sovereignty, and a prolonged period of direct conflict between Indigenous activists and the Canadian government. Flint had a "public health crisis," a loss of safe drinking water, and a public failure of law and regulation. Drawing connections between the two allows us to consider the similarities in their experiences of racialization, in their experiences within structures of imperial power, and as racialized sites of ecological ruination for those powers. As Tuck and Yang (2012) remind us, "settler colonialism is built on an entangled triad structure of settler-native-slave" (1); within their commitment to decolonization, Tuck and Yang emphasize the importance of the uncommonality of Indigenous experience. This chapter is not an attempt to recreate commonality between specific community experiences, for example by suggesting that life in Flint and Esgenoôpetitj is somehow the same (it is not), but rather to illuminate the often hidden values and structures of (post)colonial North American politics and the ways in which politics forged in the crucible of the settler-native-slave triad foment and create ruination in racialized communities for the economic benefit of so-called others. Pollution of racialized and Indigenous places in the present is not simply accidental or happenstance, but the clear result of social and political processes designed to create differential harm to such places.

Race

In his essay exploring the effect of global climate change on Indigenous communities, Kyle Powys Whyte (2017a) convincingly argues that it is not an accident or bad luck that Indigenous communities are more severely affected by climate change than many others, but a natural consequence of the racialization of global power structures. In the dispute at Esgenoôpetitj/Burnt Church, and in many such conflicts between Indigenous people and settlers ongoing in the present, many outsiders characterize isolated and impoverished Indigenous communities as solely responsible for their own poverty, and as asking for special treatment when they insist on recognition of their treaty or Aboriginal rights.

In the Burnt Church First Nation, race is not an accidental factor in the poverty and disenfranchisement of the community. The Canadian government identified people based on their racial identity as Indigenous people, and then legislated that these people must live in separate communities called reserves. These people could not borrow money based on any property they might own on-reserve, were not allowed to leave the reserve without express permission of a government agent, and were not allowed to earn money or support their families using the resources (such as fishing and forestry) that they had traditionally relied upon.

This race-based system of segregation created mass impoverishment, rendered adults unable to legally support their own children, and disrupted complex systems of relationship between people and places that had been established over millennia. When the Supreme Court of Canada recognized the treaty right of the Mi'kmaq to fish and to sell their catch to earn a moderate living, it was reinstating something that had been denied to most Mi'kmaq for generations. Only one family in Esgenoôpetitj had a commercial fishing licence and a fishing boat, because the cost of entry into the government-sanctioned fishery (over $500,000 for boat and licence) was prohibitive for most families struggling with generational poverty and unemployment.

For the Indigenous people of Esgenoôpetitj, this racialized segregation disrupted every element of their lives and livelihood, including their sense of being at home in their own place. They understand themselves as having the responsibility to maintain the well-being of, and right relationship with, the lands and waters of their traditional territories. Now they live on a smaller sliver of land beside Miramichi Bay, but they have been prevented from fully using it because their lives are governed by a race-based piece of legislation called the *Indian Act*, which prioritizes settler economies over their own. Their connection to their own waters – in terms of how they earn a living, whether they can feed their families, and when and how they can travel on the waters – is governed under race-based laws.

Flint is a deeply impoverished Midwestern city, formerly thriving under the General Motors Company, hollowed out since the auto industry's departure from the city. Like most cities in Michigan and across the US, redlining by banks, in concert with government development policies that only funded suburban developments for whites, contributed to the departure of the white middle class from Flint into the surrounding suburbs. The resulting urban fragmentation, segregation, and poverty, along with ongoing state policy, made it very difficult for cities in Michigan to thrive. These policies had similar effects in Detroit, Grand Rapids, Kalamazoo, Saginaw, and other Michigan cities, where to this day, urban centres struggle, and urban areas and school districts remain underfunded and highly segregated. Much recent data demonstrates the increasingly highly segregated nature of American cities and urban school districts (Boschma 2016); this is also borne out in the direct experience of my students from across the state of Michigan. After the Great Recession that began in 2009, many of these Michigan cities found it impossible to pay their bills, and moved closer and closer to bankruptcy. The State of Michigan, under Governor Rick Snyder, enacted a series of controversial emergency manager laws that enabled the state to supplant the democratically elected local mayor/city council government with a state-appointed emergency manager accountable only to the governor. In almost all cases where this law was used, including in Flint, a black city was targeted.

Environmental geographer and environmental justice scholar David Pellow characterizes this situation as a "democracy deficit." He argues that the problem in Flint is not actually a water problem, but a democracy problem (Washington 2016, 53–4). Pellow goes on to say that

> we live in a white supremacist society. By that I mean a society that values whiteness more than it values other communities, particularly other communities of color.
>
> In a society such as this, people of color, and in [Flint] particular people of African descent, are viewed as racially expendable. They are surplus. They are often viewed as ungovernable, as a social contaminant, a cultural pollutant. You hear how Donald Trump talks disparagingly about Mexicans who come to the United States and about Muslims – that is the language of nativism and racial expendability. People do not realize that those kinds of views are also an environmental reality. That they inform environmental policymaking or industrial practice, resulting in environmental racism. (55)

In Flint, the situation is not that it just so happened that an impoverished, largely black city experienced the destruction of its drinking water supply. Rather, after many generations of racist laws and policies targeting black communities, where they were (and are) not given the same resources, funding, democratic power, or infrastructure investment as other (majority white) cities in the same state, one of these cities experienced an ecological crisis. This crisis was not a random event, but was tied to racialized policies that have been in place over many generations, and that are creating harm in education, employment, and social well-being, just as they are in health.

Aphasia, Re-membering, and the Difficulty of Addressing (Post)Colonial Reality

For many North Americans, reflecting on and analysing their role as settlers in colonial nations[2] – and about their relationship to place in this context – is a fundamental challenge. The liberal discourse of equality often denies that racism is a systemic or everyday problem, promoting instead a "'national story' of benevolence and generosity" (Srivastava 2005, 35). Srivastava suggests that Canadians operate within "contemporary national discourses of tolerance, multiculturalism and nonracism" that mask ongoing racialized conflicts (35). Addressing the racialized structure of society is profoundly challenging because Canadian and American moral identity is so tied up in a vision of equality, a vision that, like all national visions, "requires not only sameness and communion but also forgetting difference and oppression" (Benedict Anderson, in Srivastava 2005, 39). This vision of sameness and nonracism is fundamental to the vision that the Canadian government sought to uphold in Esgenoôpetitj, and that the Michigan government used to frame its emergency manager laws.

Confronting the racism inherent in North American relationships with Indigenous peoples requires confronting fundamental questions about the history and legitimacy of the colonial states of Canada and the US. Taiaiake Alfred, an Indigenist academic, argues that

> most Settlers are in denial. They know that the foundations of their countries are corrupt, and they know that their countries are "colonial" in historical terms, but they still refuse to see and accept the fact that there can be no rhetorical transcendence and retelling of the past to make it right without making fundamental changes to their government, society, and the way they live ... To deny the truth is an essential cultural and psychological process in Settler society. (2005, 107)

Many settlers know Canada/the US as their only home, and wonder, as some of the people I interviewed in Burnt Church did, why they must pay for the sins of their forefathers. But the problems inherent in settler relationships with Indigenous peoples are not only historical; they exist in individual, social, and political lives in the present. The fundamental discomfort of reflection on race and racism makes it difficult for many to reflect upon their shared position in the colonial present.

In *Postcolonial Theory: A Critical Introduction*, Leela Gandhi (1998) emphasizes the effects of "historical amnesia" on colonial and postcolonial peoples who want to forget the trauma of their experiences and actions and move forward into a utopian future. In her essay collection *Duress: Imperial Durabilities in Our Times*, the remarkable (post)colonial historian Ann Laura Stoler explores the difficulty and inner resistance that most settlers face when addressing the unjust power relations inherent in their own colonial existences on settled Indigenous land. Stoler pushes this same early postcolonial scholarship in a new direction. She suggests that the prereflective resistance and denial that prevent conscious dialogue and engagement with colonial relationships is best understood as a form of colonial *aphasia*, in part because it is not that our knowledge of our own realities is truly lost, but that it is occluded, inaccessible, and confused (2016, 128, 166). "Aphasia is a dismembering, a difficulty in speaking, a *difficulty in generating a vocabulary that associates appropriate words and concepts to appropriate things.* Aphasia in its many forms describes a difficulty in retrieving both conceptual and lexical vocabularies and, most important, a difficulty in comprehending what is spoken" (128, emphasis in original).

Gandhi echoes Homi Bhabha's argument that recovery from the colonial experience requires a historical "re-membering, a putting together of the dismembered past to make sense of the trauma of the present" (1998, 9). Re-membering is both historical and psychological, a process of attending to the fractured experiences of the present to uncover the forgotten reality of the past. It requires us to "acknowledge the reciprocal behaviour of the two colonial partners" – that is, the mutuality of the traumatic relationship and recovery from it (Memmi, paraphrased in Gandhi, 11). For Stoler, the use of the terms "forgetting" and "amnesia" allows too much evasion of responsibility – it is not that this history is lost or forgotten, but that it is hidden, occluded, blocked off. Stoler understands the problem as a problem of aphasia because aphasics "fail to see the whole, seeing only details" (Sacks 10–11, 19, in Stoler, 166), and because, as Foucault noted, "aphasiacs dissociate resemblances and reject categories that are viable" (*Les mots et les choses*, 10, in Stoler, 167).

In her larger discussion of global (post)colonial realities, Stoler is critical of the scholarly tendency to adopt concepts as reality because of the ways in which our concepts foreclose some possibilities and overemphasize others (173–4). For example, because the concept of empire, especially colonial empires, has often precluded serious consideration of the Americas, those imperial forms are treated as anomalous and therefore overlooked. And serious scholarly examination of environmental racism and injustice in North America, even scholarship that critiques state-sanctioned violence, often criticizes capitalism without connecting it to the imperial structures that created and foment the racialized political inequities of the capitalist system. Does it matter that we recognize it as imperialism? Perhaps it does, to connect the political structures of racialized environmental ruination that are operating in the present to those that have operated in the past, and understand that present conditions are in many ways not new, but very, very old. Certainly it does, to take seriously the invitation to decolonization, and the need for the legitimate restoration of the people, lands, and waters that have become targets of the cycles of empire. Re-membering, then, is not passive recollection, but an ongoing political act of restoration, retrieving an ability to see the whole, to see categories and resemblances and resist their unconscious or prereflective rejection. Across the Americas, Indigenist activists and their allies often argue for a process of *decolonization*, a restructuring of power relationships between Indigenous people and settlers that honours Indigenous sovereignty and repatriates Indigenous land (Tuck and Yang 2012). Restoring the ability to see the whole matters for history, for decolonization, for place, for these are all ways of recognizing and seeking to protect the interconnectedness of things.

The lead pollution of the drinking water supply of the entire city of Flint, Michigan, in 2014–15, is often treated as an accident, or as an illustrative failure of one or more regulatory or planning systems, or as a critical example of the kind of environmental injustice that the American federal and state governments are charged with preventing. In Flint, the city had long purchased its drinking water from the City of Detroit Water and Sewerage Commission. To save money, Flint Emergency Manager Darnell Early planned to stop this purchase and instead draw water from the Flint River, which would be treated locally. This was the first step in a multi-year plan to build a new water system for Flint, which the elected city council agreed to. Residents were immediately mistrustful: the Flint River had long been toxic and polluted, as it had been the disposal site for effluent from the auto industry for so many years. Though the river "tested clean" in the present, their own direct experiences of pollution led locals to be wary of the water plan.

The experiences and suffering of the residents of Flint were easily dismissed by bureaucrats, over many months, even as the effects of the polluted water system became more and more apparent on the bodies of local residents. "In thinking about imperial debris and ruin," Stoler writes, "one is struck by how intuitively evocative and elusive such effects are, how easy it is to slip between metaphor and material object, between infrastructure and imagery, between remnants of matter and mind" (2016, 368). In aphasia, this slippage makes suffering and trauma unseeable, unknowable, easy to overlook, possible to explain away. The real effects of the polluted city water on the bodies of residents became impossible for the bureaucrats to see as real and material; these harms seemed to them so much more elusive than the real facts of paperwork and procedure. As a resident of another Michigan city, I would hear reports of public meetings about the water system in Flint as I drove to and from work. I would hear mothers and grandmothers talk about the effects of the water on their families' health, and hear them describe the water that was coming out of their taps. Over and over I would hear the concerns of these women dismissed – the patterns of their experiences explained away by experts who deemed the women's knowledge of their families' bodies to be irrelevant and suspect. Then one of these mothers, LeeAnne Walters, was put in touch with a lead expert at Virginia Tech, Marc Edwards (by the only EPA staffer who was speaking out about the situation, Miguel Del Torral). Edwards came to Flint and carried out independent water testing showing that the water was toxic (Lurie 2016). Dr. Mona Hanna-Attisha, a pediatrician at the hospital, released data showing widespread lead poisoning among children in Flint. With the release of these data sets, the conversation started to shift. But still, it took many weeks for so-called experts, bureaucrats, and politicians before they were willing to recognize the pattern of damage to the residents of Flint and to the city infrastructure. Now the recognition of the pattern is contained within the city. Many are willing to see that there is a pattern of pollution here, but not to recognize that this may be a part of a larger pattern of racialized disenfranchisement within the (post)colonial nations of North America that puts many more communities at risk for the same reasons.

Ruin and Ruination

Recognizing the existence and legacy of imperial processes is only one part of the necessary interrogation that must be carried out to understand the racism and imperialism that underpin our political realities. The question then becomes how imperial processes have shaped environmental realities. Are we simply dealing with the artifacts of history, the unforeseen and therefore unforeseeable consequences of power structures that were designed to create leverage over communities of people and in so doing happened to create environmental damage? When we look at the colonial ruins in our own places, it is easier to think that they happen to be there, that they were created in spite of people's best efforts, and that they persist somehow as monuments to the human spirit and ability to overcome. Perhaps this way of looking at ruins and ruination in fact illustrates (post)colonial aphasia. It is tempting to tell ourselves that this trauma happened to people and to the land by happenstance rather than through intentional structures and processes, because this alleviates responsibility. Briefly, I'd like to consider ruination (with thanks to Ann Laura Stoler for the phrase), or violence to place, in the context of race and imperial politics.

Imperial processes don't simply happen to leave behind debris, pollution, and inequity – they are designed to create and perpetuate relationships that leave some communities (both human and ecological) in ruins. Ongoing imperial processes of ruination persist in structuring places and relationships through "racialized relations of allocation and appropriation" that people must continue to wrestle with even when empires themselves cease to exist (Stoler 2016, 347). "To speak of colonial ruination," Stoler says, "is to trace the fragile and durable substance of signs, the visible and visceral senses in which the effects of empire are reactivated and remain. But ruination is more than a process that sloughs off debris as a byproduct. It is also a *political project* that lays waste to certain people, relations, and things that accumulate in specific places" (350).

The colonization of Indigenous lands and cultures across Turtle Island (North America), the transatlantic slave trade, and the industrialization of oil and gas extraction were all carefully planned, funded, and legislated projects designed to support the development of North American imperialism. In Burnt Church, the boundaries of the reserve were carefully drawn on maps and marked on roadways with red posts so that everyone driving past would know where they were. Over time, the posts became overgrown, but the boundaries were clear – well maintained roads on one side, and poor, bumpy roads on the other. Well-cared-for homes on one side, crowded homes of impoverished families on the other. These communities were created to enable the Canadian government and Canadian settlers and their corporations to carry out resource extraction on the rest of the lands and waters relatively unimpeded – to develop a forestry industry, and a fishing industry, and a mining industry, and a recreation industry without challenge or inconvenience. The ruins we face – polluted waterways, racialized communities, or falsified democracies – are carefully constructed and intimately interdependent relations of power that continue to exist and persist in doing violence to some people and some places for the sake of others' benefit. As Kyle Powys Whyte (2017a, 2017b) has convincingly demonstrated, it is not bad luck that climate injustice differentially affects Indigenous people, but déjà vu: the systematic targeting of Indigenous environmental adaptive capacity along with increasing the degree of disruptive anthropogenic environmental change that requires adaptive response. The differentiations that determine what can be ruined for wealth, and what cannot, are drawn based on race and power (not accidents or happenstance).

In his editorial in the *American Journal of Public Health*, David Rosner reminds readers of Flint's labour and environmental history, detailing the story of the strikers in the 1937 Flint March, which celebrated their ground-breaking bargaining victory in the face of violent strikebreaking efforts from General Motors (GM) and the government.

> GM not only tried to defeat its workers, but also the environment in which they, and all of us, live. The latter never even had a chance to organize and resist. Linked by roads, rivers and streams by the 1930s, Flint and the area

around it had become an industrially polluted landscape probably as bad as anywhere in the world. By 1936 the car industry had become very dependent on lead. It went into their batteries and welding, paints, lacquers, enamels and other finishes, as well as the gasoline GM cars depended on ... Huge amounts of lead and other toxins were pumped into the air, water streams and ground in and around the mammoth car factories in Flint and other Michigan cities. It is unlikely that anyone living in Flint then – or today – could escape the impact of the pollution. By the Great Depression, lead was to GM as corn was to the Midwest ... The indignities and bodily insult today's children face in Flint is horrifying. But, even more horrifying is that this city and its children have been poisoned in one way or another for at least 80 years. (2016, 200–1)

In 1937, thousands of National Guard troops were in Flint acting against strikers. In 2015, they were again in the city, moving door to door handing out bottled water. The work now is to see how these two moments are connected to one another, and to so many other moments across the continent. The confluence of laws and policies enacted by successive governments to extract resources from the land, water, and people of this place played an important role in creating this travesty.

An Environmentalism of the Built Environment

In *Thinking Like a Mall: Environmentalism after the End of Nature*, Steven Vogel challenges philosophers to develop an environmentalism of the built environment, arguing for an approach that focuses our concern on the environment *with* humans, rather than the environment *free from* humans (2015). As the stories of Flint and Esgenoôpetitj/ Burnt Church demonstrate, such an environmentalism is necessarily one of race, politics, and culture, determined to uncover the complex social and historical reality of ruination, rather than treating ruin as an accident of the past impinging upon the present. Each living person navigates this complex landscape of possibility and ruination, and the landscape itself reflects racial, ecological, and economic disparity. Because of this, there is no neutral position from which we can consider these questions: none of us can engage the problems of ecological ruination and social abandonment outside of our own bodies, without drinking water or eating food, and our ability to do these things is always already shaped by our own location within this built environment. In this (post)colonial context, the work of confronting aphasia and of re-membering place is simultaneously ecological, psychological, political, and social.

NOTES

1 With thanks to Mabel McKay and Greg Sarris, whose article "What I'm Talking About When I'm Talking About My Baskets" from *De/Colonizing the Subject: Politics of Gender in Women's Autobiography*, eds. Smith and Watson (University of Minnesota Press, 1992) illustrates the necessity and challenge of setting aside assumptions to learn what is being taught in the (post)colonial context. Thanks also to AJ Swieringa for his assistance with the research for this essay, and to Leslie Giddings for her technical support.

2 Acknowledging North American colonial history and context is central to developing an understanding of contemporary relationships (or the absence of relationships) between Indigenous people and settlers. (See, for example, Alfred 2005; Freeman 2002; Kovach 2009.) As both Indigenist and postcolonial theorists point out, the North American context is not post-colonial because in Canada and the US, the settlers have not left (Alfred 2005; Gandhi 1998). Canada and the US remain colonial nations, and settlers retain their political power and sense of legitimacy.

REFERENCES

Alfred, Taiaiake. 2005. *Wasásé: Indigenous Pathways of Action and Freedom.* Peterborough: Broadview Press.

Boschma, Janie. 2016. "Separate and Still Unequal." *The Atlantic.* 1 March. Accessed 14 August 2017. https://www.theatlantic.com/education/archive/2016/03/separate-still-unequal/471720/.

Butler, Lindsey J., Madeleine K. Scammell, and Eugene B. Benson. 2016. "The Flint, Michigan, Water Crisis: A Case Study in Regulatory Failure and Environmental Injustice." *Environmental Justice* 9 no. 4: 93–7. https://doi.org/10.1089/env.2016.0014.

Egan, Paul. 2017. "These Are the 15 People Criminally Charged in the Flint Water Crisis." *Detroit Free Press* 14 June. Accessed 28 August 2017. http://www.freep.com/story/news/local/michigan/flint-water-crisis/2017/06/14/flint-water-crisis-charges/397425001/.

Freeman, Victoria. 2002. *Distant Relations: How My Ancestors Colonized North America.* Toronto: Random House.

Gandhi, Leela. 1998. *Postcolonial Theory: A Critical Introduction.* New York: Columbia University Press.

King, Sarah J. 2014. *Fishing in Contested Waters: Place and Community in Burnt Church Esgenoôpetitj.* Toronto: University of Toronto Press.

Kovach, Margaret. 2009. *Indigenous Methodologies: Characteristics, Conversations and Contexts.* Toronto: University of Toronto Press.

Lurie, Julia. 2016. "Meet the Mom Who Helped Expose Flint's Toxic Water Nightmare." Mother Jones and the Foundation for National Progress. 21 January. Accessed 14 August 2017. http://www.motherjones.com /politics/2016/01/mother-exposed-flint-lead-contamination-water-crisis/.

Morckel, Victoria. 2017. "Why the Flint, Michigan, USA Water Crisis Is an Urban Planning Failure." *Cities: The International Journal of Urban Policy and Planning*, 62: 23–7. https://doi.org/10.1016/j.cities.2016.12.002.

Rosner, David. 2016. "Flint, Michigan: A Century of Environmental Injustice." *American Journal of Public Health* 106 no. 2: 200–1. https://doi.org/10.2105 /ajph.2015.303011.

Srivastava, Sarita. 2005. "You're Calling Me a Racist? The Moral and Emotional Regulation of Antiracism and Feminism." *Signs: Journal of Women in Culture and Society* 31 no. 1: 29–62. https:/doi.org/10.1086/432738.

Stoler, Ann Laura. 2016. *Duress: Imperial Durabilities in Our Times*. Durham: Duke University Press.

Tuck, Eve, and K. Wayne Yang. 2012. "Decolonization is Not a Metaphor." *Decolonization: Indigeneity, Education & Society* 1 no. 1: 1–40. https://jps.library .utoronto.ca/index.php/des/article/view/18630.

Vogel, Steven. 2015. *Thinking Like a Mall: Environmental Philosophy after the End of Nature*. Cambridge, MA: MIT Press.

Washington, Sylvia Hood, and David Pellow. 2016. "Water Crisis in Flint, Michigan: Interview with David Pellow, Ph.D." *Environmental Justice* 9 no. 2: 53–8. https://doi.org/10.1089/env.2016.29003.shw.

Whyte, Kyle Powys. 2017a. "The Dakota Access Pipeline, Environmental Injustice, and U.S. Colonialism." *Red Ink: An International Journal of Indigenous Literature, Arts, & Humanities*. Accessed 14 August 2017. http://kylewhyte.cal.msu .edu/s/Whyte-NoDAPL-Environmental-Injustice-US-Colonialism-4nbf.pdf/.

– 2017b. "Is it Colonial Déjà Vu? Indigenous Peoples and Climate Injustice." In *Humanities for the Environment: Integrating Knowledge, Forging New Constellations of Practice*, edited by Joni Adamson and Michael Davis. New York: Routledge.

PART THREE

Rethinking Water Policy, Practice, and Ethics

If philosophers hope to move beyond metaphysical speculation to reflect more meaningfully about lived experience of place, then that engagement, by definition, must be more than simply a theoretical exercise.

Certainly, there is scholarly value in paying heed to the significance of embodiment; of dwelling; of lived social, cultural, linguistic, and political relationships. But ultimately, the engagement between philosophy and praxis is one that opens the door to a fundamental rethinking of the role and content of public policy around water. In fact, there is a similar opportunity to rethink issues arising in the new applied field of water ethics and what it means to be engaged in responsible, ethically grounded environmental decisions and policy making.

According to some, the notion of water ethics can be traced back to the late 1970s, when discussions surrounding the normative assumptions of integrated water resource management practices arose at the United Nations Conference on Water in Argentina (Brown and Schmidt 2010, 7). By the late 1990s and 2000s, a more concerted effort to address themes of water ethics was developed through UNESCO's Commission on the Ethics of Scientific Knowledge and Technology. Fourteen documents were prepared and published in 2004 under the heading of "Water and Ethics," with other reports emerging around best ethical practices in water use over the following years (Groenfeldt 2013, 7ff). Academic conferences and texts followed, with a range of publications focused on water values, from discussions about water as a human right to the ethics of privatization of water (Whiteley, Ingram, and Perry 2008; Llamas, Martínez-Cortina, and Mukherji 2009; Brown and Schmidt 2010; Groenfeldt 2013).

In some instances, the ethical dilemmas discussed in these documents assumed an approach to ethics that Tom Beauchamp would have called "top-down" – "applying a general rule to a case that falls under the rule"

(2005, 7). The assumption in such cases is that if philosophers can provide rational moral guidelines, decision makers can simply "apply" them to the problem at hand.

Arguing that complex moral decisions require more than a simplistic application of generalized rules, principles, and theories, others suggest that ethical decision making should be "bottom-up," where moral principles become "derivative, not primary" (Beauchamp 2005, 8). Some say that "social ethics develops from a social consensus formed around cases," and that the messy process of moral decision making is an iterative process of identifying a moral *ethos* rather than a linear, deductive way forward (9).

Questioning the nature of ethics is part of a larger discussion about the future of philosophy itself. Is philosophy simply a discipline-based expert area of study, or do complex interdisciplinary societal challenges demand a new vision of the role of philosophical reflection itself? Some claim that there is a need for a complete repositioning of philosophy – one that requires a new sensibility to the public sphere and to the challenges of what Robert Frodeman calls "field philosophy" – or what, in Book VII of *The Republic*, Plato called "dwelling with the rest of the city" (2014, 85).

What this ongoing dialogue – as well as this section of the book – points to is that ethical decision making, as an embodied, implaced process of framing questions, appropriating traditions, and informing praxis, is, at the very least, *complex*. **Bryan E. Bannon** responds to the complexity of decision making about stream daylighting: not by advocating for a set of abstract moral rules but by reminding us of the need to engage in a deeper relationship of what he calls an ecologically restorative "friendship" with the natural environment.

Trish Glazebrook and **Jeff Gessas** focus on issues of water protection, specifically in light of the effects of a pipeline on Lake Oahe. Questioning modernist paradigms, the authors show how phenomenological, Indigenous, and ecofeminist perspectives help to inform wiser, morally robust strategies for decision making.

Henry Dicks then helps us to understand how the development of water policy itself assumes a certain taken-for-granted understanding of the *polis*. Decision making is less a matter of developing a rational theoretical model and then *applying* it to complex water issues than it is a challenge of unpacking hidden assumptions that affect how we frame water policies within the context of the polis.

Ultimately, as **Bob Mugerauer** shows us, moral decision making arises from within an ethic of care and critical awareness of the complexity of embodied, lived networks of relations. It is, as philosopher Joseph

Kockelmans reminds us, never wise to "separate school from life" (1983, 381). It is never wise to separate prescriptive moral discourse from the embeddedness of social, political, regulatory, economic, technological, and ecological dimensions of place.

REFERENCES

Beauchamp, Tom. 2005. "The Nature of Applied Ethics." In *A Companion to Applied Ethics*, edited by R.G. Frey and Christopher Heath Wellman. Malden, MA: Blackwell Publishing, 1–16.

Brown, Peter G., and Jeremy J. Schmidt, eds. 2010. *Water Ethics: Foundational Readings for Students and Professionals*. Washington: Island Press.

Frodeman, Robert. 2014. *Sustainable Knowledge: A Theory of Interdisciplinarity*. New York: Palgrave Macmillan.

Groenfeldt, David. 2013. *Water Ethics: A Values Approach to Solving the Water Crisis*. New York: Routledge.

Kockelmans, Joseph J. 1983. "The Foundations of Morality and the Human Sciences." In *Foundations of Morality, High Rights, and the Human Sciences*, edited by Anna-Teresa Tymieniecka and Calvin O. Schrag. Dordrecht: D. Reidel Publishing Company.

Llamas, M. Ramón, Luis Martínez-Cortina, and Aditi Mukherji. 2009. *Water Ethics*. New York: Taylor & Francis.

Whiteley, John M., Helen Ingram, and Richard Warren Perry. 2008. *Water, Place and Equity*. Cambridge, MA: The MIT Press.

9 The Bonding Properties of Water: Community, Urban River Restoration, and Non-human Agency

BRYAN E. BANNON

The farmhouse lingers, though averse to square
With the new city street it has to wear
A number in. But what about the brook
That held the house as in an elbow-crook?
I ask as one who knew the brook, its strength
And impulse, having dipped a finger length
And made it leap my knuckle, having tossed
A flower to try its currents where they crossed.
The meadow grass could be cemented down
From growing under pavements of a town;
The apple trees be sent to hearth-stone flame.
Is water wood to serve a brook the same?
How else dispose of an immortal force
No longer needed? Staunch it at its source
With cinder loads dumped down? The brook was thrown
Deep in a sewer dungeon under stone
In fetid darkness still to live and run –
And all for nothing it had ever done
Except forget to go in fear perhaps.
No one would know except for ancient maps
That such a brook ran water. But I wonder
If from its being kept forever under,
The thoughts may not have risen that so keep
This new-built city from both work and sleep.
                                 – Robert Frost (1923), "A Brook in the City"

Frost's poem posits the disappearance of nature from our lives as a source of the malaise that underlies contemporary urban life. Although the situation and the poem are both more complex than can be reduced to such a simple idea, the fate of the brook he describes is a common one in the history of urban development in the United States: for purposes of controlling floods, increasing the amount of land for development, or carrying away industrial waste more efficiently, headwater streams are frequently buried, channellized, or filled. The US has never treated its rivers particularly well, precipitating such infamous events as the Cuyahoga River fire of 1969, which was neither the first time the Cuyahoga burned, nor the worst of such fires. The economic development of US cities has been tied to the fate of their rivers, but this has come at steep ecological cost.

As we become increasingly aware of the myriad ways human development affects the health and integrity of riparian systems, particularly in urban areas, a movement has emerged oriented not only towards cleaning up urban rivers, but also towards reversing the course of the development Frost describes by unearthing these headwater streams, exposing them once more to the light and air. This form of stream restoration, colloquially referred to as daylighting, can have large ecological and economic effects in urban areas. As with other restoration projects, lingering ethical and practical concerns accompany daylighting, including whether the costs are warranted given other social needs such as education, health care, or even other environmental remediation projects; whether the project can be successful and by what measures such success will be assessed; whether and how best to include community input regarding the project; how exactly to restore the system – either through more sustainable channelization or to a so-called natural state – and at what scale restoration is necessary; etc. Regardless, that municipalities, community members, and restorationists are negotiating these issues seems to signal a desire to reform the relationship between urban settlements and their sustaining waters.

Nevertheless, these are only signals of a détente, and there are good reasons to believe that certain ways of carrying out such restoration projects will embody the human domination of nature in much the same way as burying or channelizing the streams did in the first place. Given that, as Ellen Wohl et al. (2005) point out, in many cases river restoration projects are undertaken without good ecological modelling of the river, a clear understanding of ecosystem function, the spatial and temporal scale of the project itself, or even any long-term assessment of a project's success, it is clear that, at least in those cases, restoration work can be undertaken for reasons other than conservation. These motivations, however noble some may be, are where the domination of nature creeps back in for, as these authors note, "human expectations of ecosystems reflect the prevailing culture and views of how ecosystem operation is related to quality of life." As many environmental scholars have argued from a variety of perspectives, currently prevailing cultural views about nature lie at the root of destructive environmental practices. In this way, restoration projects frequently embody the same social prejudices – such as the subordination of ecological systems to human temporal scales (Bannon 2014) – that reproduce nature's domination.

The potential problem of domination does not lie in the fact that there are ineliminable social dimensions to restoration projects, however. On the contrary, many environmentalists have noted that ecological restoration practices possess the potential to rebuild relationships between humanity and the rest of nature.[1] Acknowledging the social dimensions of restoration, it becomes necessary not only to align restoration projects with our best scientific knowledge, but also to grapple with the social meanings of water, the political challenges of urban sustainability projects, and the humanism that usually underlies such discussions. Although the aim of this paper is not to resolve these issues, I offer here an alternative way of approaching them by employing a phenomenological theory of meaning and value. In what follows, I will argue that such a framework can enrich our understanding of the social domain to be inclusive of the non-human co-creators of the natural world and, in so doing, make some suggestions regarding how the concept of friendship can be helpful in the practice of river restoration within cities.

The argument will proceed through three phases. First, I examine more closely the deeper underlying problems of social meaning that surround water policy in general, focusing on what Vandana Shiva (2002) and David Groenfeldt (2017) call the cultural water wars. Since both situate the source of conflicts over water in the domain of meaning, I introduce in the second section the phenomenological theory in which meaning and value are said to inhere in relations rather than be a property of the thing in question. Relational views are helpful to environmental practices because they exemplify more fully ecological ways of thinking. When applied to rivers and their restoration, this theory illuminates the possibility of viewing the river as an agent, albeit not of the traditional anthropological model. In the final section, keeping in view the agency of the river, I will advocate friendship as a model capable of attuning restorationists to a variety of values that can establish a more democratic approach to their practice while still not compromising their commitment to a practice guided by science. Central to all these considerations is the question of environmental identity: Who are we as individuals, as human beings, and how do our relationships with our environs shape those identities?

Riparian Restoration and Modernist Thinking

Urban stream restoration has a variety of benefits for the human community in which the restoration occurs, for the myriad forms of life that inhabit riparian systems, and for the continued existence of the river itself. In fact, stream restoration is usually regarded as undoing the kinds of damages that unchecked human development has wrought within the landscape. As Matthew Gandy notes, recent times have even seen a movement away from the technocratic practices that have historically characterized urban development towards more communal, democratic, and ecologically informed practices (2014, 16–17). Moreover, advocates of restoration have argued that for this very reason, restoration practices can build bridges between communities and landscapes that will serve to prevent future degradation of the landscape (e.g., Light 2000). After all, if one has participated in the shaping of the land itself, the thought is that one will become invested in the fate of that place. Restoration, then, appears as a promising practice in mending the human relationship to the rest of nature. Consequently, it is unclear how such a practice would contribute to the further domination of nature.

The seeds of domination, however, lie dormant in many soils, one of which is the alluvium of meaning. Socially taken for granted meanings frequently determine which practices are undertaken and how those practices are executed, and practices of ecological restoration are no exception. As R. Bruce Hull and David P. Robertson (2000) indicate, how practitioners of ecological restoration frame their practice in language can have large effects on the decisions made about restoration goals. Additionally, human beings generally understand themselves, their situation in the world, and the relationships they enter into through identities shaped by social meaning. Changes in environmental behaviours – and hence support for ecological restoration – are at least in part dependent upon perceived values (Ives and Kendal 2014), which are in turn dependent upon how human beings conceive of their identities and consequently of what an appropriate relationship with the rest of nature should be like. It proves necessary, then, to look at the meanings of water as expressed in ecological restoration projects to ensure that the practices they encourage and the relationships they create do not promote the further domination of nature or human groups.

One potential worry emerges from the discourse around "water wars." The term typically refers either to conflicts between human groups over the development of water resources or to actual armed conflicts over access to water. Yet, as Vandana Shiva (2002) explains, even as these are very real problems, we are also engaged in a completely distinct second water war, one that influences the human relationship to water in general and has the potential to diminish both human autonomy and nature itself. This water war is not so much over the control of water resources, but over the *meaning* of water and the consequent shifts in our relationships to it. Shiva describes this less acknowledged water war as "a clash between two cultures: a culture that sees water as sacred and treats its provision as a duty for the preservation of life and another that sees water as a commodity, and its ownership and trade as fundamental corporate rights" (2002, x). On this account, the conflict does not *only* exist at the level of who gets access to water and how water is to be used, but also at the level of the value of water in human life itself: Is it merely a commodity, or is it something more? On the surface, conflicts over water rights can be understood as a straightforward competition for access to a resource, but what is really at stake is cultural survival, which hinges on sustaining certain meaningful relations to water. The right to water emerges from "a given ecological context of human existence" (20), meaning that the ways in which cultures have developed practices for inhabiting a given area allow them to live within the limits of available water. These practices serve to define the culture itself, so when current international policy enforced through institutions such as the International Monetary Fund and the World Bank undermine such practices, water policy is destructive of culture itself (cf. chs. 4 and 6, especially). This kind of water policy impedes what Kyle Powys Whyte calls the "collective continuance" of a culture, a culture's capability to adapt its practices to changing external social and natural forces (2016, 166). The goal of resisting such policy is not to remain unchanged, but to maintain the ability to adapt on one's own terms rather than on the terms of the dominant group.

David Groenfeldt (2017) chronicles the effects the water wars have upon the collective continuance of cultures in the American West by means of economic coercion into participating in development projects that go against established cultural practices. He describes the situation aptly in saying, "cultural water wars and sub-wars (skirmishes) can revolve around almost anything – dams, rivers, institutions – because it is not the thing as a thing, but the thing as a repository and communicator of cultural meaning that has the capacity to motivate cultural conflict" (133). According to Groenfeldt, the primary means by which this war is waged is by "expropriating the meaning and values that local communities traditionally confer on water and water ecosystems," a process Groenfeldt attributes to "semiotic hegemony" (130). These hegemonic power relations are such that individuals or groups may feel they have no options beyond those options afforded to them within a dominant group's belief system, and for this reason they peaceably adopt the views of the dominant group (127–8). In so adopting the hegemonic views of the dominant culture, these individuals and groups abandon their relationships to the landscape as well, resulting in the disappearance of what made that culture unique: a particular relationship to the surrounding world.

Although urban riparian restoration is unlikely to result in the destruction of a particular cultural relationship to the land, what these accounts reveal as a danger is the way practices, meaning, and the relationship to the natural world are interconnected. Following from these ideas, the question I pose is whether predominant approaches to urban riparian restoration embody the prejudices and values of the dominant society that have led to our current environmental situation and thereby contribute to the continued domination of the landscape or whether they reflect the transformative practice that some restoration advocates suggest. The answer, unsurprisingly, is ambiguous: though I agree with the advocates of restoration that the practice has the potential to be transformative, it is hard to ignore the way in which, in seeking public support for restoration projects, practitioners typically emphasize humanistic values and prejudices stemming from Western dualistic metaphysics, by which I mean the strong separation of natural and social values.

The trouble is that riparian restoration is an amazingly diverse practice undertaken for a variety of reasons, ranging from the improvement of animal habitat, erosion control, flood plain reconnection, enhancing recreational opportunities, etc. For that reason, it is impossible to make blanket proclamations regarding the motivations and priorities of restorationists in general. Perusing the literature, approaches to ecological restoration can be grouped into two dominant paradigms in the United States: one oriented towards the restoration of ecological integrity, defined largely in terms of how ecological systems functioned before the intervention of European settlers, and the other oriented towards the provisioning of so-called ecosystem services.

The first, more ecocentric paradigm situates the value of restoration in returning ecological integrity to an area rather than upon establishing ways for humans to inhabit their larger environment. This paradigm tends to focus on past ecological functioning, which is understandable given how river restoration projects in general "require an understanding of natural systems at or beyond our current knowledge and present a significant challenge to river scientists" (Wohl et al. 2005). Practically, this approach in the riparian context has tended to focus on restoring historical forms (e.g., redirecting the flow of a river) or restoring processes (e.g., flooding regimes) (Wohl et al. 2015). Emily Bernhardt et al. (2005, 636) describe the most common goals for river restoration projects as "(i) to enhance water quality, (ii) to manage riparian zones, (iii) to improve in-stream habitat, (iv) for fish passage, and (v) for bank stabilization." Advocates of this approach tend to rely on definitions of the natural in contradistinction to the artificial to establish what the goals for a productive restoration are, even if they accept that naturalness is a spectrum.[2] These approaches tend to view human activity and alterations of the environment as anathema to the naturalness of a place, which raises problems in the context of the cultural water wars: Even as these goals are laudable, they are defined independently of cultural relationships to the river and tend to regard the river as a patient on which restorationists act.

The second paradigm is more problematic in the context of the water wars and more fully embodies the kinds of values Shiva and Groenfeldt criticize. Mark Everard and Helen Moggeridge (2012) epitomize this approach based in ecosystem services. In their view, the purpose of ecological restoration is straightforwardly anthropocentric: to promote "socially-beneficial river functions." Drawing from the UN Millennium Ecosystem Assessment, they identify four categories of services: "'provisioning services' (extractable resources), 'regulatory services' (processes that regulate the natural environment), 'cultural services' (culturally-valued benefits), and 'supporting services' (processes essential to maintenance of the integrity, resilience and functioning of ecosystems)" (296–7). They distance themselves from those seeking to restore rivers for the sake of "biodiversity-concern" (310), citing instead how the ecosystem services concept "provides a means to quantify benefits to society" (297) by means of monetization, which allegedly allows the use of markets in favour of environmental conservation efforts. This paradigm flagrantly considers water as a commodity like any other, subject to privatization and trade, wise use or exploitation, but those aspects of the human relationship to water that fall outside of the quantifiable do not even play a role in the concept. While these flaws may not be inherent in the ecosystem services concept itself (Muraca 2016), this framing of the concept lends itself to perpetuating the water wars rather than ending them. If both paradigms embody problematic tendencies of Western culture, what is to be done?

The problem with restoration in the context of the water wars is not so much with the practice itself as with particular ways in which it is shaped by the dominant Western dualism of nature and culture. To put an end to these cultural water wars that underlie and enable the more visible conflicts over water, this dualism needs to be overcome. To this end, Shiva recommends viewing water as "sacred" (2002, 131–9). Doing so entails nothing supernatural, but simply recognizing that "the worth of water rests on its role and function as a life-force for animals, plants, and ecosystems" (137). Contrasting the sacred with the utilitarian, Jerome Delli Priscoli (1998, 635) offers a congruent sense of the term, defining it as "those aspects of water through which mystery and unknown, or some would say irrational, elements become present to our awareness." Mystery, in the phenomenological tradition, refers less to the irrational, and more to the inexhaustible depths of ways in which a being can appear. Taken this way, preserving the sacredness of water requires us to examine ways in which water reveals itself in the world through its relations and thereby creates a panoply of meanings.

If the deeper concern in the water wars is a hermeneutical one regarding the very meaning of water, and if the resolution of the cultural water wars requires an increased attentiveness to the myriad ways in which water manifests meaning and value, then the next step in developing a better restoration practice surrounding rivers will be to examine how a relational theory of value, as is offered in phenomenology, can help deconstruct the semiotic hegemony that leads to the domination of water. I will argue there are two consequences to adopting such a theory in the context of river restoration: 1) that thinking about meaning and value in non-dualistic terms can provide a foundation for ecological restoration; 2) that it becomes necessary to engage with the river itself as an agent in restoration practices, thus leading us to concerns beyond the human.

Meaning, Value, and Agency

If the majority of stream restorations do not escape from the hermeneutical context that enabled the domination of the river in the first place, it becomes necessary to displace that way of thinking about the river itself, about water, and about ourselves so as to institute a different kind of relationship between ourselves and water. Here is where phenomenology can help: where the majority of ecocentric positions in environmental ethics predicate themselves upon the establishment of some form of intrinsic value to the natural world, phenomenological positions understand meaning and value as being intimately related and seek them in relations rather than in individual beings.[3] This shift in understanding can be useful in the context of river restoration because it broadens the focus of our concern beyond the human and allows there to be values other than the economic and technologically oriented ones dominant in Western cultures.

In the ecological sense, nature is not a thing itself, but a contingent whole comprised of the relations between various beings. From this perspective, we can speak of nature at various levels, whether we are speaking of it on a global scale, wherein "nature" refers to the totality of global ecological relations, or on a local scale, wherein it refers to the relations between the specific living and non-living constituents of a place. On this account, humans are natural, and our works do not destroy naturalness but transform its internal relations, which may be for the better or the worse depending on which standards one deploys in making the evaluation. What no longer makes sense, however, is to say that human presence necessarily degrades a landscape: Our activity does not destroy the naturalness of the landscape, but we might criticize certain ways of interacting with the landscape and encourage others. The question becomes how to make such an assessment.

One possibility is to analyse the meaning *in* a landscape, but to do so from a perspective of embodied meaning rather than viewing meaning as a property of reason or formal languages. As embodied beings, we are able to affect other bodies and to have ourselves shaped in turn by those bodies with which we interact. Because they affect us, these bodies matter to us in various ways; they come to embody particular meanings depending on how they affect us, and the value of those meanings becomes enshrined through habitual practices. "Spicy," for example, is a particular way my body responds to consuming a hot pepper. The foods I typically consumed as a child predisposed me to avoid spicy foods and consider them exotic and perhaps hazardous to my health. These meanings and values evolve and adapt over time in light of new experiences. After having positive experiences with spicy foods, I might rehabituate myself to enjoy the sensations that accompany them or I might come to appreciate the effects such foods have on my body. Of note, however, is how meaning and value of this kind are not the sole purview of human beings, but of bodies in general: all bodies affect others and are affected by other bodies. It is easy to see how this is so in the case of animals, since their vital situations establish meaningful worlds around them through their bodily organization and behaviour. Consider how spiciness is present to other animals: lacking the bodily ability to be affected in that way, birds have a different relationship to hot peppers than I do. I can understand the meaningful experience of these other bodies by observing how various animals construct meaning in the world around them with our corporeal similarities aiding in the translation. So when I see the squirrel recoil after eating the birdseed I have mixed with cayenne in the birdfeeder and then avoid that seed entirely, I see that it too has some relationship to spiciness and that meaning has a value to it. As bodies become increasingly foreign, it becomes increasingly difficult to understand their experience. I have, for example, little understanding of the meaningful relations of certain plants, but, even so, through interacting with them I can gain an understanding of their behaviours and the value certain other bodies (e.g., bees or beetles) have in their lives.

The claim that meaning and value inhere in bodily relations is as true of inanimate bodies such as stones as of other living beings. Although certain experiential concepts make no sense to attribute to inanimate beings – pain isn't part of a stone's bodily capabilities – we can still speak about certain meaningful relations that we observe. In a sense, this is what we are doing when we discuss physical interdependence. Since no two things ever relate to each other in isolation but their relation always takes place on the background of other relationships, meaning and value operate on three levels: 1) the products of the relationship between bodies, 2) the way in which the context of the relationship shapes the relationship itself, and 3) the effects those relationships have on the larger system of relationships. In other words, the meaning of a specific phenomenon for another body is relative to how that phenomenon affects the body, to how environmental conditions affect that relationship, and to how the relationship shapes environmental conditions going forward. Meaning, then, will largely be descriptive of how specific relationships create predispositions for bodies to behave in particular ways within environments. Bruno Latour (1994) calls these material predispositions "programs of action": bodies are predisposed to interact with other bodies to produce certain effects, but these programs are frequently redirected, disrupted, enhanced, etc. through their interactions with other bodies in ways that bring about novel effects.

How value inheres in relations for non-conscious bodies is less apparent. In conscious bodies, value emerges from ends the body takes up in navigating the world, but non-conscious bodies seem to lack vital situations that would make such values possible. We can take a clue from Latour on this point, however. If non-conscious bodies possess programs of action, are open to affection, and exert influence over other bodies to shape environmental conditions, then value can be discovered in how such activities enhance or detract from the array of possibilities present for the body and within the larger environment. Although it may not be possible to describe with precision what value is within specific relations, in general positive values for the body in question enhance that body's potential for action without significantly degrading its environment, whereas negative values diminish that potential or degrade the environment to inhibit long-term development.

A river is an interesting phenomenon from this perspective because considering the river as a being requires us to synthesize all of these factors, since it comprises an assemblage of biotic and abiotic relations. The river is the geological terrain and the water that flows through it, for sure. But it is also the vital activity of the multitude of species that affects the flow, shape, and integrity of the riverbed and that in turn relies upon the water for various reasons. The river, taken in this way, embodies a diversity of programs of action, including such things as nutrient and sediment transfer and providing habitat. Each program, however, is inherently relational: every program of action embodies a disposition for the river behaving in a particular way in encountering the agency of another being, which is itself shaped by its experience of the river and a variety of other beings. The agency of the river lies in the kinds of effects it engenders, how it shapes the agency of others, the manner in which it shapes the environmental conditions in which it exists. This agency, though not conscious, constitutes the practices and habits that engender meaning and value in the world.

This way of understanding meaning and value within riparian systems can be helpful in resolving the potential problems discussed above with respect to urban river restoration. If the river itself can be considered as an agent, a non-conscious being with its own meaningful relations and particular values, restoring it will require more than seeking to reinstate particular functions of the river, specific historical assemblages of species, or any one set of conditions that it may have displayed in the past. Rather, restoring the river involves both these ecological considerations *and* establishing the kind of human relationships to the river that will serve to preserve and develop the various programs of action present for the river. Just as Aldo Leopold encouraged us to "think like a mountain" to improve the human relationship to the mountain's ecology, so too we can think like a river in an effort to determine how best to live without diminishing the potential of the river. In the last section of this paper, I will argue that friendship can provide one model for thinking about how to relate to the river and how restoration can become an expression of that friendship.

Friendship and Rivers

The restoration of urban rivers is not typically thought to be a priority in ecological terms. Wohl (2005) expresses the reason: "The potential for ecological improvement is limited, and the principal benefits from a restoration project are social, such as building a sense of community by involving residents of a neighbourhood or increasing pride in place." For this reason, the motivations behind restoring urban rivers tend to be more economic than ecological, even as certain ecological goals are integrated (e.g., fish passages). If one defines naturalness in terms of the absence of humanity, surely the notion that ecological improvement is limited is true. After all, Boston and Providence are not going anywhere, so restoring the Charles or the Providence Rivers to a precolonization state is simply not feasible. But if we think about humans as being a part of and co-creators of their environments – as all animals are – then it becomes less clear why restoring urban rivers cannot in fact be extremely ecologically relevant. Could current attempts to reconnect the greenbelt around Boston, Frederick Law Olmstead's "Emerald Necklace," which involves unearthing headwater streams of the Charles such as the Muddy River in the process, be a step towards making the Charles a more ecologically vibrant place?[4] Could the Providence River waterfront be restored ecologically as well as economically and culturally (Martin 2010)? The challenge is to create a healthy urban ecology that integrates what are typically thought of as social and natural factors. In this section, I will trace how the theory of meaning and value developed in the previous section can help us perform this integration by thinking about the human relationship with the river in terms of a friendship.

Friendship is a relationship best understood in terms of a project of mutual co-shaping.[5] Although it could be argued that *every* relation serves to define the being in question, the first defining dimension of friendships is the openness to being shaped by the other within the relationship. As Marilyn Friedman (1999) notes, friendships, unlike familial or other community relationships, are particularly meaningful and powerful in the life of a person because they are *chosen*. For this reason, friendships display one's own ideals for living more clearly than other interpersonal relationships. Hence, a second defining dimension of a friendship is a commitment to the friend such that we choose to include its projects among our own. Taking up specific projects for and with the friend is an expression of one's openness to being shaped by the friend and results in developing new habits and practices. Since habits and practices establish meaning and values in the world, friendships contribute to how we come to think about ourselves and the world around us. A third defining dimension of friendship, then, lies in how it concretizes a distinct identity for the parties by establishing a set of practices through which we engage with the world, others, and ourselves. Attempting to live up to the identities refined within the friendship does not always go as expected: We make mistakes, misinterpret, and occasionally do things to harm our friends. If these occasions become habitual, they become good reason to rescind the friendship. More typically, however, friendships are characterized by a willingness to forgive the other party and repair the relationship. This tendency towards maintaining the relationship through practices of forgiveness is a fourth defining dimension of the relationship. To be friends with a river is therefore to choose to make oneself open to being shaped by one's relationship to the river through the practice of taking up projects on behalf of and with the river, including making amends for past mistreatment.[6]

It is easy to see that humanity could take up this kind of relationship towards the river, but it is less clear that the river could reciprocate. Despite the fact that a river does not consciously promote or detract from the well-being of the individual human beings engaging with them, it is equally undeniable that it possesses the kind of agency described in the previous section: it embodies specific programs of action that are shaped by both its constituent relations and its relations to its environs (e.g., flooding). Because programs of action are relational, the agency of the river works upon us, influencing our thoughts, feelings, and behaviours as readily as our human agency can work on the river. In this way, the river's programs of action, that is, its agency, can be oriented towards enhancing or detracting from human agency and well-being. Moreover, these relations are inherently meaningful, allowing us to interpret the agency of the river and impute value to the changes that occur as a result of our interactions with the river. Even as we seem to be interpreting the river's activity as we would a person's speech, that analogy fails to grasp the opacity of the river and our inability to attribute conscious awareness or intentionality behind the river's activity. Behaviour, in the robust sense of agency's responsiveness to situational factors, does not require consciousness. The river, then, can be said to reciprocate our friendly gestures towards it, though we would still be mistaken to attribute volitional choice to the river in its reciprocation.

Viewing riparian restorations as an ongoing expression of a relationship between human inhabitants and the river is not without precedent. Sally Eden, Sylvia Tunstall, and Susan Tapsell (2000) use a case study of the restoration of the River Cole, a tributary of the Thames, to demonstrate the promise of looking beyond the nature–culture dualism dominant within Western cultures in favour of the sort of model espoused here, in which the river is viewed as an agent. Colleen Fox et al. (2017) offer a number of examples of restoration projects involving Indigenous communities, several of which involve conceptions of the river as a living being with whom we have a relationship, wherein the knowledge of the river contributed through the inclusion of the Indigenous groups significantly improved the project and social relationships between settler groups and Indigenous nations. In all these cases, the acknowledgement of the agency of the river and consequent formulation of restoration plans that seek to accommodate humanity as a participant in the life of the river and a part of its ecology improve the quality of the restoration through achieving a higher level of engagement from human participants.

In the urban context, being a friend to a river can prove difficult for a number of reasons, ranging from the hiddenness of headwater streams in culverts, to the density of settlement and accessibility of the river, to the availability of opportunities to do so, etc. Especially given that waterfront property can be privately owned, require cleanup of industrial pollutants, or already be developed with hard engineering features such as concrete channels, flood barriers, dams, etc., finding occasions to undertake restorations can be challenging. These conditions make certain activities impossible (e.g., afforestation or restoring a flood plain), of course, but do not preclude others. The Charles River Watershed Association, for example, sponsors volunteer canoe trips to remove the non-native water chestnut (*Trapa natans*) choking portions of the Charles throughout the greater Boston area. The felicitously named group Friends of the Los Angeles River has also had success lobbying for the restoration of parts of that river despite development pressure and much of the river being hard engineered into, essentially, a giant drainage ditch to prevent flooding in the city (Gottlieb and Azuma 2007). The group has been raising awareness regarding the ecological possibilities of the river at their "Frog Spot" education centre, and a number of artists have been exploring alternative understandings of the space from the Western technocratic one embodied in its concrete channel (Gandy 2014). So even though engaging with the river can be difficult in urban contexts, it is possible to use such activities to imagine and implement alternative ecological futures in which human activities are integrated into restoration programs to create a positive ecological future in the process. Such re-envisioning of the interaction between humanity and the landscape represents one way in which the hermeneutical water wars over the river can be brought to a détente.

The challenge of using friendship as a guide for restoration is that there is no one set of prescriptions that one can offer, since the contours of friendships are defined by the parties that enter into them. However, as the parties within the friendship come to develop together, the relationship itself comes to possess its own immanent norms and practices because of the commitment to the relationship itself. Because friendship demands that one engage in practices that promote the flourishing of the other party, the relational understanding of meaning and value becomes critical to humanity engaging with the river in a friendly manner. The river may never be able to *speak*, but meaning is present in countless ways in the relationships that constitute the river, and we can, from these meanings, begin to impute value to certain relations and practices insofar as they affect those relations.

Attempting to understand the meaning and value our practices have within the context of friendship can be a productive model for rethinking restoration practices for a number of reasons, but foremost is the manner in which it provides responses to some of the vexing questions about the goals of restoration. When a new practice engenders a greater variety of possibilities than were present previously, without sacrificing or compromising present capabilities, that practice marks an improvement in ecological conditions. The goal of urban restorations, then, should not be to restore ecological function as if humanity had never been present. Such a goal encourages us to think of urban ecologies in a dualistic manner that can entrench anthropocentrism as a central value of restoration practice. After all, if cities are considered to be sites of ecological ruin from the outset, any restoration ought to be oriented towards what is good for humanity because the place is already conceived of as a strictly human place. Rather than put ourselves in this position, the goals of urban restoration practices should include enhancing ecological possibilities, promoting environmental justice by creating healthier environments for all citizens and not merely aesthetically pleasing water features for wealthier citizens, and creating opportunities to engage citizens in preserving their rivers to strengthen the bonds of community between them. Human beings can and must be a part of transforming their urban spaces in a way that achieves this heightening of ecological possibilities in a just and equitable manner. Restoration practices could be part of a larger urban renewal; the river gives back to us.

Aside from providing a reason to engage in ecological restoration along urban rivers, another benefit of friendship for urban riparian restoration is the flexibility of thinking in terms of what restoration entails. Although it is good to employ historical fidelity for restoration so as to stay within the limits of human understanding of ecological function and thereby restore ecological integrity, doing so is not sufficient, as Eric Higgs (2003) has argued; restoration work also calls for what he refers to as "focal practices." Focal practices are activities that *create* meanings that help us make sense of our world and place within it rather than *reproduce* patterns of meaning that preserve the fragmentation and dualistic tendencies of modern Western life. The result is "wild design," wherein part of the design of the project is to leave open what the restored place will look like, how it will behave, how it will be used, etc. In the urban context, there is much promise here as well: restoring waterfronts to bring about engagement with the river rather than passive enjoyment of it, subordinating economic development to ecological ends, making space for people to continue to participate in the life of the river through volunteer work, festivals, community planning ... all of these practices contribute to building stronger bonds between people and place.

Conclusion

Speaking of the Los Angeles River, filmmaker Wim Wenders observes, "landscapes ask for their own stories to be told. The L.A. River, as it now exists as a cemented river, has a story of aggression to tell" (quoted in Gottlieb and Azuma 2007, 41n1). This story of aggression is not found in meanings a subject imposes upon a passive nature, but is rather the present and observable relations bespeaking a history of domination, a history of pernicious anthropocentrism in which the river was subdued and subordinated for human ends. Restoring the river involves telling a different story without being able to erase what we have already done, which requires transforming existing relations, establishing new meanings in the landscape, and leaving room for a diversity of values within our places. Thinking about restoration, and our environmental activities in general, in terms of relationships is a good step in that direction.

The trouble is that the environmental narratives that surround us are extremely complex and frequently difficult to read, especially when a landscape is considered to be the product of a multitude of agencies with their own distinct programs of action. In any given situation, there is likely no course of action that will result in unequivocally positive results for all parties. But, in encountering these situations, it is important not to succumb either to appealing to historical nature alone to resolve the difficulty or to thinking of ourselves as separate from our local ecologies or somehow contaminating pristine nature with our very presence. As I have proposed here, friendship can provide some guidance on how to act in such cases and how to respond to the consequences of acting.

But the larger issue at stake is the manner in which considering our relationship with the river to be a friendship extends beyond restoring the non-human ecology of the river. The water wars we have been concerned with are conducted at the level of both dominating practices and dominating conceptual frameworks and their ravages negatively affect both humans and the rest of nature. If, as Shiva and Groenfeldt make clear, certain hegemonic relationships to water can destroy the collective continuance of cultures by making the practices that define them impossible, it is equally true that by establishing new practices and norms surrounding the cohabitation of the river, new cultures can emerge. These new cultures, which restoration can make possible in urban spaces, promise to change the tenor of the Anthropocene narrative:

rather than a story about how humanity despoiled the planet, we might begin to help our terrestrial friends develop new capabilities. Although we cannot return to the house in the elbow-crook of the brook that Frost nostalgically describes in his poem, liberating the immortal force of the brook by unearthing it can help us to avoid the sleepless nights by discovering more meaningful work dedicated to improving the living systems on which our own well-being depends.

NOTES

1 Wohl et al. (2005) admit that "in many urban areas ... the potential for ecological improvement is limited, and the principal benefits from a restoration project are social, such as building a sense of community by involving residents of a neighbourhood or increasing pride in place." Such claims are consistent with ones made by Andrew Light (2000; 2002)and Eric Higgs (1991).
2 See Angemeier (2000) and Hettinger (2005) for two different approaches in this vein.
3 Intrinsic value theories in environmental ethics tend to take objectivist or subjectivist forms. The former argues that intrinsic value exists in virtue of some objective property of the environment, and the latter argues that the value exists in virtue of a subject valuing the environment for its own sake rather than for some instrumental value. Katie McShane (2007; 2009) offers several excellent typographies of the concept and, in my view, the strongest defence of it. For a more detailed account of why I believe intrinsic value theories are unnecessary in environmental ethics than what is offered here, see Bannon (2013).
4 For information on this project, see Rosen (2017) and Muddy River Restoration Project (2017).
5 I offer a more detailed account of both why friendship is important and how it could operate in an environmental context in Bannon (2017).
6 There are, of course, problematic dimensions of friendship as well, and I do not want to pretend they do not exist. Simply because they exert such powerful force in the shaping of our character and identity does not require that they provide the best interpretations of the world, orient us towards acceptable ends, or otherwise make us decent people. For this reason, I take friendship to be an *ethical* practice rather than a *moral* one. That is, the practice of friendship cultivates a specific identity through the establishment of habits, meanings, and values, but there is no inherent orientation towards virtuous activity. It may result in desirable, pro-social tendencies such as increased empathy, but even those tendencies may be limited in scope (e.g., only towards one's friends) or establish in-group/out-group dynamics that

can create problems. These concerns need to be addressed, but it would be too great a digression to address them here.

REFERENCES

Angemeier, Paul L. 2000. "The Natural Imperative for Biological Conservation." *Conservation Biology* 14 no. 2: 373–81. https://doi.org/10.1046/j.1523-1739 .2000.98362.x.

Bannon, Bryan E. 2013. "From Intrinsic Value to Compassion: A Place-Based Ethic." *Environmental Ethics* 35 no. 3: 259–78. https://doi.org/10.5840 /enviroethics201335325.

– 2014. "Resisting the Domination of Nature: Regarding Time as an Ethical Concept." *Environmental Philosophy* 11 no. 2: 333–58. https:// doi.org/10.5840/envirophil2014101317.

– 2017. "Being a Friend to Nature: Environmental Virtues and Ethical Ideals." *Ethics, Policy & Environment* 20 no. 1: 44–58. https://doi.org/10.1080 /21550085.2017.1291824 .

Bernhardt, Emily S., Margaret A. Palmer, J.D. Allan, G. Alexander, Katie Barnas et al. 2005. "Synthesizing US River Restoration Efforts." *Science* 308 no. 5722: 636–7. https://doi.org/10.1126/science.1109769.

Delli Priscolli, Jerome. 1998. "Water and Civilization: Using History to Reframe Water Policy Debates and to Build a New Ecological Realism." *Water Policy* 1 no. 6: 623–36. https://doi.org/10.1016/s1366-7017(99)00019-7.

Eden, Sally, Sylvia M. Tunstall, and Susan M. Tapsell. (2000). "Translating Nature: River Restoration as Nature-Culture." *Environment and Planning D: Society and Space* 18 no. 2: 257–73. https://doi.org/10.1068/d180257.

Everard, Mark, and Helen L. Moggridge. 2012. "Rediscovering the Value of Urban Rivers." *Urban Ecosystems* 15 no. 2: 293–314. https://doi.org/10.1007 /s11252-011-0174-7.

Fox, Coleen A., Nicholas James Reo, Dale A. Turner, JoAnne Cook, Frank Dituri, et al. 2017. "'The River Is Us; The River Is in Our Veins': Re-Defining River Restoration in Three Indigenous Communities." *Sustainability Science* 11 no. 3:521–33 online. https://doi.org/10.1007/s11625-016-0421-1.

Friedman, Marilyn. 1989. "Feminism and Modern Friendship: Dislocating the Community." *Ethics* 99 no. 2: 275–90. https://doi.org/10.1086/293066.

Frost, Robert. 1923. *New Hampshire, a Poem; with Notes and Grace Notes.* Project Gutenberg ebook. Accessed 3 March 2019. https://www.gutenberg.org /files/58611/58611-h/58611-h.htm.

Gandy, Matthew. 2014. *The Fabric of Space: Water, Modernity, and the Urban Imagination.* Cambridge, MA: MIT Press.

Gottlieb, Robert, and Andrea Misako Azuma. 2007. "Bankside Los Angeles." In *Rivertown*, edited by Paul Stanton Kibel. 23–46. Cambridge, MA: MIT Press.

Groenfeldt, David. 2017. "Cultural Water Wars: Power and Hegemony in the Semiotics of Water." In *The Politics of Fresh Water: Access, Conflict, and Identity*, edited by Catherine M. Ashcraft and Tamar Mayer. New York: Earthscan.

Hettinger, Ned. 2005. "Respect for Nature's Autonomy in Relationship with Humanity." In *Recognizing the Autonomy of Nature*, edited by Thomas Heyd. 86–98. New York: Columbia University Press.

Higgs, Eric. 1991. "A Quantity of Engaging Work to be Done: Ecological Restoration and Morality in a Technological Culture." *Restoration & Management Notes* 9: 97–104. https://doi.org/10.3368/er.9.2.97.

– 2003. *Nature by Design*. Cambridge, MA: MIT Press.

Hull, R. Bruce, and David P. Robertson. 2000. "The Language of Nature Matters: We Need a More Public Ecology." In *Restoring Nature: Perspectives from the Social Sciences and Humanities*, edited by Paul H. Gobster and R. Bruce Hull. 97–118. Washington, D.C.: Island Press.

Ives, Christopher D., and Dave Kendal. 2014. "The Role of Social Values in the Management of Ecological Systems." *Journal of Environmental Management* 144: 67–72. https://doi.org/10.1016/j.jenvman.2014.05.013.

Latour, Bruno. 1994. "On Technical Mediation: Philosophy, Sociology, Genealogy." *Common Knowledge* 42: 29–64.

Light, Andrew. 2000. "Ecological Restoration and the Culture of Nature: A Pragmatic Perspective." In *Restoring Nature: Perspectives from the Social Sciences and Humanities*, edited by Paul Gobster and R. Bruce Hull. 49–70. Washington, D.C.: Island Press.

– 2002. "Restoring Ecological Citizenship." In *Democracy and the Claims of Nature*, edited by Ben Minteer and Bron P. Taylor. 153–72. Lanham, MD: Rowman & Littlefield.

Martin, Laura Jane. 2010. "Reclamation and Reconciliation: Land-use History, Ecosystem Service, and the Providence River." *Urban Ecosystems* 13 no. 2: 243–53. https://doi.org/10.1007/s11252-009-0110-2.

McShane, Katie. 2009. "Environmental Ethics: An Overview." *Philosophy Compass* 4 no.3: 407–20. https://doi.org/10.1111/j.1747-9991.2009.00206.x.

– 2007. "Why Environmental Ethics Shouldn't Give up on Intrinsic Value." *Environmental Ethics* 29 no. 1: 43–61. https://doi.org/10.5840/enviroethics200729128.

Muddy River Restoration Project (MRRP). 2017. "Muddy River Restoration Project Overview." Accessed 21 May 2017. http://www.muddyrivermmoc.org/restoraton-overview/.

Muraca, Barbara. 2016. "Re-appropriating the Ecosystem Services Concept for a Decolonization of 'Nature.'" In *Nature and Experience: Phenomenology and the Environment*, edited by Bryan E. Bannon. 143–55. New York: Rowman & Littlefield International.

Rosen, Andy. 2017. "Stringing the Emerald Necklace Back Together." *The Boston Globe*, 16 April.

Shiva, Vandana. 2002. *Water Wars: Privatization, Pollution, and Profit.* Cambridge, MA: South End Press.

Whyte, Kyle Powys. 2016. "Indigenous Experience, Environmental Justice and Settler Colonialism." In *Nature and Experience: Phenomenology and the Environment*, edited by Bryan E. Bannon. 157–73. New York: Rowman & Littlefield International.

Wohl, Ellen, Paul L. Angemeier, Brian Bledsoe, G. Mathias Kondolf, Larry MacDonnell, et al. 2005. "River Restoration." *Water Resources Research* 41 no. 10. https://doi.org/10.1029/2005WR003985

Wohl, Ellen, Stuart N. Lane, and Andrew C. Wilcox. 2015. "The Science and Practice of River Restoration." *Water Resources Research* 51 no. 8: 5974–96. https://doi.org/10.1002/2014wr016874.

10 Standing Rock: Water Protectors in a Time of Failed Policy

TRISH GLAZEBROOK AND JEFF GESSAS

Mni Wiconi: Water is life. Water is sacred.

Plans to build the Dakota Access Pipeline (DAPL) across 1,900 kilometres from North Dakota to Illinois, at a cost of approximately USD $3.7 billion, were announced in 2014 (Businesswire). The pipeline passes under Lake Oahe, the Standing Rock Reservation's water source half a mile from the reservation. The Lakota Sioux living on the reservation mobilized to protect this water through peaceful, prayerful direct action and established the Oceti Sakowin camp. By mid-September, it hosted over a hundred tribes and thousands of others at what is thought to be the largest gathering of Indigenous Americans in over 150 years (Tannous 2016), and one of the largest Native American protests ever (Mother Jones 2017). We argue that this event is a game-changing, historic rupture of the logic of capital by a phenomenology of water.

Our approach is ecofeminist in several ways. First, we assume that exploitation of nature and political oppression are inherently linked by history, theory, praxes, and policy (Warren 1996, x–xvi). Second, our assessment of the oil industry uses Warren's theoretical concept of logics of domination: i.e., oppressive conceptual frameworks that establish a dualism, privilege one term over the other, and then use that privilege to justify exploitation of the latter by the former (1996, 21–4). In this case, the dualism is man/nature, but man/woman is also discussed below, insofar as the logic of capital critiqued as oppressive is patriarchal. Third, we use an ecofeminist methodology of narrative voice by examining first-hand accounts of water protectors and drawing from Lakota teachings and practices evidenced in activities, interviews, social media commentaries, and other web-based sources. We use these sources because mass news media are largely co-opted by the logic of capital and corporate interests (Herman and Chomsky 2002). But also, listening prioritizes relationships, and reveals ethical attitudes and behaviours often overlooked in mainstream Western ethical theory (Warren 1996, 27). Voice is especially important in cross-cultural research, where imposed meaning can overdetermine the other's experience. Irigaray (1985) diagnoses such imposition as a "logic of the same" that judges the other according to one's own standards, against which the other will always appear lesser. For example, in canine terms, cats are deficient retrievers. We aim not to speak for the Lakota, but to listen and learn about the Lakota Way of Life so that others living in mainstream consumer culture, for whom events at the Oceti Sakowin camp may appear as just another environmental action, can also see how Lakota being-in-the-world is fundamentally different from the destructive logic of capital manifest in DAPL's attitudes and practices towards nature and people.

Our ecofeminist approach is also a phenomenology. As a term used technically in this phenomenology, "logic" carries epistemological but also ontological force. Epistemologically, "logic" means a way of thinking; ontologically, it means how things appear – that is, what they are – in that way of thinking. The phenomenological task at hand is to witness the way of thinking and ontology of the Lakota Way of Life, and to understand water in that logic. Heidegger argues that "'phenomenology' means ... to let that which shows itself be seen from itself in the very way in which it shows itself from itself" (1986, 34; 1962, 58), and later argues for "releasement toward things" that does not view things reductively through a projected paradigm (1992, 23; 1966, 54). Releasement rather "lets one wait" (1995, 226) so beings have opportunity to speak for themselves. In Heidegger's analysis, modernity projects a technological paradigm that reduces nature to mere resources, and "the world becomes without healing, unholy" (1994, 295; 1971, 117). Poets, he says, "bring to mortals the trace of the fugitive gods, the track into the dark of the world's night" (1994, 319; 1971, 141) that is modern technology's assault upon nature. We argue that the water protectors of Standing Rock open the possibility of a broader cultural healing. Our ecofeminist phenomenology aims to make visible in the Lakota Way of Life "the possibility of dwelling in the world in a totally different way" (Heidegger 1992, 23; 1966, 54) from the Western, Eurocentric, technoscientific, militarized culture of capital.

First, we briefly show that US water policy is a failed system. Second, we provide an overview of the history and logic of technoscientific capital. Next, we show how the oil industry, in particular DAPL, epitomizes this logic. Then we show how the Lakota Way of Life entails a completely different way of thinking. In conclusion, we make policy recommendations based on our analysis towards functional policy.

US Water Policy

Water policy in the United States is convoluted: rather than being managed by a single, comprehensive policy, water is subject to federal, state, county, and municipal management. Policies differ depending on water's location: for example, groundwater from aquifers differs legally from spring and surface water. Rainwater is governed by different rules based on geographical distinctions; for example, in mountainous regions of Colorado, rainfall collection is regulated to maintain rivers for downstream users, but this is not the case in the hills of Texas that drain into aquifers. Legal access and ownership vary according to political and geological factors; a plethora of precedents in some places still use the settler policy of prior appropriation – that is, the first non-Indigenous person to access the water owns it. Water may be in the public domain or owned privately. In most states, rights can be bought, sold, and leased. In some areas, use requires a legal permit (Samson 2008). Federal policy originally managed water quality and system flow across state and other boundaries, and left allocation decisions to state and territorial governments (Getches 2001). The Reagan administration moved water policy and decision making largely to the states, who developed a watershed management approach in which governance bodies in the drain area for a body of water manage the system collaboratively (Gerlak 2006). Water policy and management remain collaborative across government levels and regions, and tribal governance has an uneven voice. For example, fifteen Indigenous tribes participate in the Columbia River Inter-Tribal Fish Commission to renegotiate the Columbia River Treaty between the US and Canada (Columbia 2017), but elsewhere, American Indian tribes are not as actively engaged in policy intervention.

Disputes are common, and resolutions are not always consistent. Policy and infrastructure are fragmented in ways that make flexibility and adaptive approaches difficult. Policy makers face challenges in supply and in ongoing damage to water quality. They balance increasing demand against aquifer decline, and satisfy competing interests between municipalities needing clean drinking water, tribes relying on fish as a nutritional staple, farmers relying economically on irrigation, and industrial use. Agricultural runoff pollutes water with animal waste, hormones, and chemical pesticides and fertilizers. Urban storm water runoff contains toxins from car exhaust. Industrial pollution is a substantial health threat. In 2010 alone, industrial facilities dumped over

100 million kilograms of toxic chemicals, polluting 22,500 kilometres of rivers and 8,900 hectares of lakes, ponds, and estuaries, into 1,400 US waterways in fifty states (Kerth and Vinyard 2012). The City of Flint, Michigan, has been in acute water crisis since the 1970s because of bacterial, industrial, agricultural, and lead contamination for which the city and state, as well as the polluting corporations, take little, if any, responsibility (CNN 2017). A class action suit resulted in USD $17 million in damages in 2017 and the arrest of five state officials, including the head of the Health Department, on charges of involuntary manslaughter for covering up lead contamination (Atkinson and Davey 2017). While pollution is common, accountability is rare. These water access and quality issues are amplified by climate impacts of changing rainfall patterns and catastrophic events like drought and flood, for which federal policy is sorely lacking and local adaptation responses are limited by governance attitudes and funding constraints.

In the absence of functional policy, oil is a focal threat to water. A break in an Enbridge Energy pipeline in 2010 put nearly 4,000,000 litres of crude oil in the Kalamazoo River, which remained closed into 2012 (National Transportation Safety Board 2010). Despite knowing about the break in the pipeline, the operating firm delayed shutdown for 17 hours. Enbridge was fined USD $3.7 million (National Transportation Safety Board 2012). Against its 2016 adjusted earnings of approximately USD $4,662 million (Enbridge 2017c), however, a fine of less than 0.1 per cent of earnings provides no meaningful accountability. On its website, Enbridge claims to include "environmental performance goals into our business systems" (Enbridge 2017a). It has been on the Global 100 Most Sustainable Corporations index for eight years in a row, and was ranked twelfth worldwide on *Newsweek*'s 2016 Green Rankings corporate sustainability index (Enbridge 2017b). Corporate and industry image management manufactures a kind of cognitive dissonance between public image and the reality of impacts.

In short, water policy in the US is a failed system in crisis that is fragmentary and driven by needs management and economic factors, rather than long-term vision, transparency, and accountability. Interpreting policy in law is unpredictable and appears capricious. The system is easily manipulated. For example, in applying for permits to build the pipeline, DAPL treated the project as a series of small construction sites eligible under the Nationwide Permit 12 (NWP 12) process that exempts projects from the environmental review required by the *National Environmental Policy Act* and the *Clean Water Act* (Moyers 2016). When oil development and transport put water at risk, policy offers little protection. Legal remedies at best provide post hoc compensation that reduces ecosystem and human health to an exchange value. Even if life and well-being were morally exchangeable in this way, profits dwarf the value assigned to living beings.

Big Oil and the Logic of Capital

The logic that trades the health and well-being of people, non-human others, and ecosystems for profit is the logic of capital. Fossil fuels play a central role in this logic because modern culture requires an energy source to drive its exploitations of nature. An "unreasonable demand" is thus "made that [nature] supply energy that can be extracted and stored" (Heidegger 1997a, 18; 1977, 14). This section shows how "Big Oil" paradigmatically embodies capital logic as a logic of domination that reduces nature to a mere resource and denigrates people, cultures, and alternative knowledge systems. First, the section provides a history of the logic of capital through an overview and development of Heidegger's assessment of the intellectual history of the West (Glazebrook 2000). Next, the section provides international examples of ecosystems' destruction and abuse of people by multinational oil companies, and follows this by demonstrating that the Canadian and US governments are mobilized to promote and protect oil industry interests while overpowering the democratic processes of resistance. The next section contrasts the phenomenologic of life demonstrated by the Lakota at Oceti Sakowin against this logic of capital.

Aristotle explained natural entities as things that grow and develop of their own accord on the basis of an inner teleology: for example, puppies become dogs, acorns become oak trees. In contrast, Aristotle held that artifacts are created by a human artist who envisions them in advance based on their intended purpose (1941, 640a32; cf. 1140a13). Aristotle's word for artifact was *technē*, which is the etymological origin of "technology." In the medieval Judeo-Christian account, nature becomes a divinely created artifact: what for Aristotle was driven by inner purpose is now driven by divine intent. In modern science, however, God is – as LaPlace is reported to have told Napoleon – an unnecessary hypothesis (Ball 2003, 388). Nature has been stripped first of its own purpose and then of divine intent, rendering it nothing more than bodies in motion in a mechanistic, materialist universe. Accordingly, it is readily available in modernity to be appropriated to human needs. Modern technology thus "sets upon nature" (Heidegger 1997a, 18; 1977, 15), which is no longer understood as what grows of its own accord, but as material for human production. Science and technology have collapsed into "technoscience," that experiences things as nothing more than resources for human consumption (Heidegger 1997a, 36; 1977, 32).

Technoscience mobilizes "the organized global conquest of the earth" (Heidegger 1997b, 358; 1982, 248) as a logic of domination based on the dualism of "man" and "nature." In this dualism, "man" is the master rather than the member of an ecosystem. Human mastery is explicitly established at the roots of modern science in Bacon's intent to "conquer nature in action" (1980, 16, 21). Bacon's approach is also misogynistically violent, as his experimental method intends to extract conclusions "out of the very bowels" (1980, 23) of nature when "she" is "under constraint and vexed ... [and] forced out of her natural state" (1980, 27). "She" is removed from "her" natural state in the ideal conditions of the laboratory that afford the precision of calculation. Galileo's universe is accordingly "a book written in the language of mathematics" (1957, 238). This quantification prepares nature to be exploited in the exchange value of capital (Glazebrook and Story 2015, 131–2).

In this logic of human mastery, not all humans are, however, equally privileged: colonialism aligns human others with nature, reducible also to resources, in slavery, for example, or simply disposable, in the neo-colonial politics of oil development. In Nigeria, the "resource curse" of oil development led to corruption and violence against the peoples of the Niger Delta, who had access to less than 5 per cent of the USD $340 billion in oil revenues generated in the Delta in the last four decades of the twentieth century (Gary and Karl 2003, 25). As their land, water, agriculture, fishing, and health were destroyed by leaks from pipelines routed by multinational oil companies wherever was most convenient – sometimes right through people's homes – the people organized resistance. Government response, allegedly supported and financed by oil companies, included violent strategies of assault, rape, imprisonment, and execution (Glazebrook and Kola-Olusanya 2011, 165). Likewise, the genocide in Darfur, Sudan, was not only financed by oil revenues but also enabled by practical support from oil companies; for example, the planes and helicopters that firebombed villages gassed up at fuel stations and flew from airstrips owned by the Canadian corporation Talisman Oil (Glazebrook and Story 2012). Globally, oil development is a violent assault with virtually no accountability for the destruction and suffering it causes.

The oil industry is also driven in the global North by interests of capital, with little if any accountability for the irreversible impacts of pollution in ecosystems, and the longer-term climate impacts of burning fossil fuels. Climate change is already causing a massive humanitarian crisis in hunger in Africa, for example (Glazebrook 2011; Lott et al. 2013). Canada and the US are better equipped to respond, but unable to avoid climate impacts (Lemmen and Warren 2004, ix–xxiv; Karl et al. 2009, 12).

A well-established regulatory system in Canada (Glazebrook 2003, 28–30) disallows practices such as gas-flaring, common in Nigeria, which releases toxins into the atmosphere, causing a wide variety of human health issues and cutting life expectancy (Glazebrook and Kola-Olusanya 2011, 174). Despite what is seemingly a culture of accountability and regulation, Canadian Prime Minister Justin Trudeau supports the development of the northern Alberta oil sands and the construction of three new pipelines, including the transnational Keystone XL Pipeline to carry oil from the oil sands to Nebraska, despite protests on both sides of the border. Tim MacMillan, head of the Canadian Association of Petroleum Producers, warns of the "poor economics" that will result if the pipelines are not built, and emphasizes the need for "new oil export pipelines" in "new and growing markets and our existing very large markets," while saying nothing about risks and impacts (Bakx 2017) – despite at least 750 metric tons of oil spilt in Canada in 2015 alone (List of oil spills 2017). Neither government nor non-governmental industry associations have shown they can be trusted to protect ecosystems and future generations from the environmental costs of oil.

In the US, oil producers have been building pipelines to transport oil across the country since Edwin Drake drilled the first commercial well in Pennsylvania in 1859. According to industry sources, liquid petroleum is transported across the US through a 306,000-kiilometre network of thirty-four pipelines (Pipeline 101 2016; ITA 2017). Hundreds of metric tons of oil are spilt from pipelines in the US every year (List of pipeline accidents 2017). The oil industry is possibly the single most influential voice in US governance. The infamous 2010 Supreme Court ruling on *Citizens United* enabled corporations to fund election campaigns and influence voters (*Citizens United v. FEC* 2010). The Koch brothers, currently the richest oil magnates in the US, support a network that allegedly spent USD $750 million in the run-up to the 2016 election (Vogel and Johnson 2016). The current president is surrounded by Koch allies. Frayda Levin, board chair for the main voter mobilization group of the Koch network, is confident in the network's power through "relationships we built over the years with senators and congressmen" (Vogel and Johnson 2016). Big Oil whispers loudly in the federal government's ear.

At the other end of the political spectrum, the citizens of Denton, Texas banned hydraulic fracturing – "fracking" – in city limits by a landslide vote in November 2014 (Fry et al. 2015, 100). Within hours, the City was sued by the Texas Oil and Gas Association and the state's General Land Office (Fry et al. 2015, 105; Heinkel-Wolfe 2014). In May 2015, Texas Governor Greg Abbott signed the "ban on bans" bill that denies cities the right to regulate or limit oil and gas development (Buchele 2015).

As a result, Denton was obliged to lift its ban. The industry mobilized state government to protect its interests, and overpowered democratic process aimed at protecting a community from health and other risks of fracking.

In summary, the capital logic of Big Oil is a violent logic of mastery and domination that reduces nature to resources with only exchange value, prioritizes profit over human and ecosystem well-being, readily violates human rights, and overwhelms democratic process. The next section seeks healing.

The Lakota Way of Life and DAPL

This section examines the Lakota Way of Life in contrast to DAPL's logic of capital. The foundational difference is that the logic of capital first and foremost values money. The Lakota, however, have still not claimed a 1980 settlement of over USD $1 billion awarded as compensation for US appropriation of the Black Hills of South Dakota in 1877 (Justia 2017, 372). The settlement threatens tribal honour and unity, and accepting it would mean giving up the claim on the Hills (Frommer 2001; Streshinsky 2011). These attitudes towards money are so fundamentally different that the two ways of thinking are incommensurable – there is no common ground that could measure both without submitting one to the terms of the other in a projective "logic of the same." Yet something different is needed in the face of the inability of water policy to protect water from the oil industry, as well as broader policy inadequacy to protect the environment, human health, biodiversity, and everything else oil harms. As noted on the Defund DAPL web page, the contemporary "financial system is outdated. At its heart lies a blind devotion to short-term profit at all costs" (Defund DAPLa). We use four axes of difference to provide a way to understand these two different ways of being-in-the-world as a gesture towards healing: peaceful praxis and violence; respect and disregard for other knowledge systems; mastery and membership; and authority and self-governance.

Peaceful Praxis and Violence

The violence of DAPL includes the assault on the people gathered at the Oceti Sakowin camp, as well as their vilification. During the Labour Day weekend of 2016, DAPL, aware of a pending injunction to prevent digging on land suspected of being a sacred burial ground, brought in bulldozers overnight and cut a 3.2-kilometre, 4.5-metre-wide trench through the area. Water protectors attempted to halt the digging by chaining themselves to equipment, but DAPL brought in security forces, including Frost Kennels of Ohio; they pepper-sprayed protestors and attacked them with dogs, who bit six water protectors and a horse. When a line of Indigenous men linked arms and walked towards the dogs and their handlers chanting, "We are not afraid," the handlers quickly retreated to their vehicles with the animals. Frost Kennels later admitted they were not licensed to use dogs. Members of the press were arrested and charged with criminal trespass, or felony conspiracy with a maximum sentence of forty-five years (Mother Jones 2017).

Subsequently, water cannons were used in sub-zero temperatures. Clothes could be heard "crunching" as they froze on people's bodies. The Healer Council fruitlessly demanded an immediate halt to the use of water cannons because of "the real risk of loss of life due to severe hypothermia under these conditions" (Dakwar 2016). Rubber bullets were fired directly at people's heads (Koerth-Baker 2016); these maimed permanently when used at close range (Iron Eyes 2017). Angela Bibens (2016) of the Indigenous Environmental Network reported on one 48-hour period in which tear gas, pepper spray, and mace resulted in permanent blindness to one woman and the cardiac arrest of an Elder, while many others suffered seizures and vomiting; water cannons were mixed with mace and fired at non-participant recorders; and a sound cannon was fired at the on-site medical station. A volunteer required twenty surgeries to repair her arm after being struck by a concussion grenade. These assaults were documented by the American Civil Liberties Union and the Indigenous Environmental Network, and recognized as violations of human rights by the United Nations and Amnesty International (Bibens 2016; Dakwar 2016; Iron Eyes 2017). In December, DAPL hired TigerSwan (May 2016), a private security firm of counter-terrorism experts run by retired US Delta Force members with ties to Blackwater, a mercenary organization notorious for its work in Pakistan and Iraq (Scahill 2010). Video evidence widely posted on social media shows water protectors engaged in peaceful, prayerful action even when facing continuing assault by this "privatized military force" (Levin 2017).

To justify their violent response to the water protectors, the Morton County Sheriff's Department described them as "very aggressive," and their demonstration as an "ongoing riot" (Dakwar 2016). This was coupled with more subtle public undermining. For example, Energy Transfer Partners, DAPL's parent company, officially holds that the Standing Rock Sioux Tribe had had many opportunities since 2014 to provide input on the pipeline's route (Stevens 2016). In fact, the Standing Rock Tribal Council voiced its objection to the pipeline's routing under their water supply right away in 2014 in a Tribal Council meeting with DAPL representatives (Standing Rock Sioux Tribe 2014). Their objections made no difference.

DAPL, the Morton County Sheriff's Office, and mainstream media continued to call the water protectors "protestors." This term projects DAPL's ontology of violence onto the water protectors by reducing them to "us–them" confrontation. Furthermore, it obscures their goal, otherwise clear in "water protectors." As "protestors," they are separated from their goal and reduced to a vague existence that obscures relating to them with ethical concern, just as "meat" disrupts the consumer's potential for an ethical relation towards the animal in industrialized food systems. "Protestor" thus carries "empty intention" in the Husserlian sense: that is, content, meaning, and experience are not filled in, though such fulfillment is required for meaningful knowledge (Husserl 2001, 274). Hearing about "protestors" dehumanizes because it does not provide a connection to the lived experience of knowing that water is at risk, and to understanding why and how it should be protected. "Protestor" rhetoric therefore undermines public support for the water protectors. These words are neither translations nor interchangeable, because they name the people at the Oceti Sakowin camp from out of incommensurable ontologies that function in capital's logic of domination on one hand, and a logic of respectful caring, community, and healing on the other.

Donald Trump, first and foremost a businessman functioning in the logic of capital, signed an Executive Order to restart pipeline construction as one of his first acts in office. He waived the fourteen-day notice to Congress, which Amnesty International called "an unlawful and appalling violation of human rights" (Archambault 2017). Construction began immediately. In February, the Oceti Sakowin camp was cleared by security forces armed with machine guns and tanks. The section under Lake Oahe was completed in March. Oil began flowing through the pipeline on 1 June. State security costs were estimated above USD $22 million (Nicholson 2017), several million over North Dakota's planned budget. These costs include paying for local law enforcement, who made some 800 arrests over the course of the action; DAPL is presumably responsible for TigerSwan's fees that were incurred privately. North Dakota has asked for federal support to help it cover its shortfall; at least some of the cost for the violence against water protectors is falling to taxpayers. Democratic process includes rights to assembly and free speech. The violent response to the water protectors demonstrates that the oil industry is not limited by democratic process, and that public funds can be mobilized to support corporate interests over the public interest of access to unpolluted water.

In contrast to this violent logic of domination, the water protectors' ontology affords strong care for water. The primary expression of the action is *Mni Wiconi*: Water is life; water is sacred. It is much more than a resource for human use. Bobbi Jean Three Legs explains the spiritual significance of water as "first medicine" that "goes back to being a mother. Your baby is first coming from water, so it's very sacred. And your babies are in water for nine months before they even breathe their first breath of air" (Three Legs 2017). Dr Cheryl Crazy Bull further explains the strength of water as medicine: "the blood of First Creation, *Inyan*, covers *Unci Maka*, our grandmother earth, and this blood, which is blue is *mni*, water, and *mahpiya*, the sky. *Mni Wiconi*, water is life" (Crazy Bull 2016). Water constitutes life's origin, sustains life, and provides the medium for life's renewal in the next generation.

This account of the sacred nature of water speaks to the Lakota connection of the living to ancestors and future generations. "The elderly are our history. They are sacred," says the Lakota woman Sápa in an interview, and also, "the children are sacred because they are our future" (Gambrell 2016). Black Elk – edited, albeit controversially, by John G. Neihardt – described the "nation's hoop," the traditional, circular arrangement of tipis that is "a nest of many nests, where the Great Spirit meant for us to hatch our children" (Neihardt 2008, 156). The circular hoop, which, unlike a straight line, has no privileged beginning and end point, echoes the cycle of life over generations. Ancestors are present in the now by safeguarding life for the next generation, and the current generation carries the responsibility of honouring that safeguarding and themselves pass it forward to the future.

This intergenerational relationship is at the heart of other Indigenous American traditions, for example, the Iroquois Seventh Generation principle: "In every decision, be it personal, government or corporate, we must consider how it will affect seven generations into the future" (Larkin 2013). The principle is said to have originated in Iroquois governance practices as far back as the twelfth century, articulated by the Great Peacemaker Deganawidah in the Great Law of Peace (The Great Peacemakers). It exhorts decision-makers to "think of the continuing generations of our families, think of our grandchildren and of those yet unborn, whose faces are coming from beneath the ground" (The Great Peacemakers), and is "one of the first mandates given us as chiefs, to make sure and to make every decision that we make relate to the welfare and well-being of the seventh generation to come ... What about the seventh generation? Where are you taking them? What will they have?" (Vecsey and Venables 1994, 173–4).

The sense of community extends beyond the Lakota themselves and includes caring for others and respecting them across chasm-like cultural difference. Multiple videos witness water protectors telling security personnel that they understand they are doing a job and are not bad people. This serves the practical goal of humanizing both themselves and the security personnel, perhaps strategically interrupting the latter's capacity to treat water protectors violently. Even if strategic, this approach also realizes an a priori logic in which security personnel are fully human in their capacity for choice-making and reasoned thinking, and acknowledges this personhood directly to them rather than reducing them to a tool of DAPL. For those who came to the Oceti Sakowin camp, one water protector described it, after months of assault by security forces, as a prayerful place of refuge that is meaningful and helpful for everyone, and that the sense of community on the ground is stronger than the security presence with helicopters and armed officers on hilltops surrounding the camp (Grady 2016).

The depth of the healing capacity of this logic of respect, caring, and community was further demonstrated in December 2016, when a group of veterans (estimates vary wildly from about 500 to 2,200) travelled to Standing Rock to support the water protectors. They came from all walks of life across the country, and included a 97-year-old woman who had served as a nurse in the Second World War (Petersen 2016). The vets came to provide support, but once they arrived, something else happened that culminated in a formal ceremony. The vets apologized to the Sioux for the army's role in the oppression and genocide of Native Americans. Virtually all the vets present were strongly moved by this ceremony of forgiveness. Despite the continuing threat of the pipeline being routed under the Standing Rock Reservation's water supply, and months of brutal abuse by security, the water protectors' seemingly inexhaustible capacity for care provided support and healing to ex-members of a military that has a history of brutally oppressing Native Americans in general – including the Lakota specifically in the 1890 massacre at Wounded Knee in which hundreds of Lakota Ghost Dancers, many of them women and children, were slaughtered.

A significant difference between the knowledge systems at work here is the experience and conception of time. The logic of capital works in terms of monthly, quarterly, and annual budgets. Elected officials in government think as far forward as the next election. The Iroquois Seventh Generation principle mentioned above expresses a commitment to intergenerational justice. The water protectors often cite providing future generations with what they need to survive as a driving motivation. The logic of capital simply does not have enough capacity to grasp the time extension needed to devise functional policy that can sustain the contemporary resource base for even the current generation. The water protectors' ontology functions in a conception of the real that has the capacity to think in terms of the time frames necessary for sustainability. One hundred and twenty years after Wounded Knee, water protectors bring a functional ontology of time to failed US water policy.

Respect and Disregard for Other Knowledge Systems

In Vandana Shiva's critique of the science-based knowledge system of the global North, other knowledge systems are given no respect (1991, 34–5). For Heidegger, likewise, technoscience "drives out every other possibility of revealing" – that is, understanding nature (1997, 31; 1977, 27). Across the planet, knowledge learned and practices developed over generations are trivialized, ignored, or dismissed as myth, superstition, or, because the logic of capital is patriarchal and so devalues on the basis of gender, "old wives' tales." The logic of capital accepts only the technoscientific knowledge system that enables its global exploitation of ecosystems and people.

Those immersed in this logic, such as those working in the oil industry, security and law enforcement, and governments, are blinded by the ontology of capital that simply does not afford vision of alternative knowledge systems. This does not mean that there can be no accountability for acts of brutality or law-breaking, and it is unlikely that oppressors are so incapable of self-reflection that they have no understanding at all of the injustices they enact, even if their logic provides justifications they take as mitigating. But it does mean that the logic of capital entails blindness to anything outside and irreducible to its logic of the same. Any disruption of this logic and its obliviousness will require a holistic system shift that cannot be accomplished easily, akin to the shift from geocentrism or belief that the earth is flat.

The Lakota Way of Life, however, as a marginalized perspective, does not have the luxury of taking its knowledge system as the only functional truth, given its embedding in the hegemonic logic of capital. History obliges the Lakota to understand that logic as the broader context that situates the Lakota Way of Life. When Kem Gambrell interviewed four Lakota women in positions of leadership, they attributed their success to having a strong connection to the Lakota Way of Life, achieving education beyond high school, and being able to understand and operate within the mainstream, non-Lakota world (Gambrell 2016). The "Lakota Way of Life" refers not only to traditional, cultural, and spiritual practices, but also an ethic of collectivity and relationality that prioritizes connections through community, education, and crossing cultural boundaries. Education is especially important, because it is not only how children grow into the Lakota Way of Life, but also how adults are able to negotiate the neocolonial contexts that continue to appropriate and oppress their lived experience. After settler colonization, in the final decades of the nineteenth century, educating Lakota children into the technoscientific, neocolonial knowledge system was obligatory. Under a residential school program, some 30,000 children were removed from their families into a program of forced enculturation and "Christianization" that attempted to discipline away their language, clothing traditions, and culture based on the assumption that "any Indian custom was, per se, objectionable, whereas the customs of the whites were the ways of civilization" (Kneale 1950).

The women interviewed by Gambrell understand bicultural education as a significant empowerment strategy. Education in the Lakota Way of Life teaches the meaning of *Mni Wiconi* to support understanding of the effect of not treating water as merely a resource (Crazy Bull 2016). At the same time, assimilation into mainstream education allows Lakota to "educate scientists, environmental resource management specialists, and fisheries and wildlife managers ... science teachers, environmental lawyers, and economists who study ways to sustainably manage natural resources on Lakota homelands" (Crazy Bull 2016). Bicultural education affords introduction of the Lakota Way of Life into mainstream education, while at the same time equipping Lakota for employment in resource management offices on Lakota homelands. Moreover, sending the youth off-reservation to college educates "the next generation of political leaders in tribal leadership, government studies, and law so that they may work on behalf of their communities," and provides tools to manage all aspects of Lakota land, "which is our lifeblood, and the source of our lives" (Crazy Bull 2016). Bicultural education allows youth to be assimilated into US culture, economics, policy, and governance on Lakota terms, rather than the terms of the logic of capital that is constantly undermining the resource base.

The capacity to blend cultures and languages is also important. Many water protectors, in interviews and in communication with the USACE and courts, move back and forth between Lakota and American English. This mixture of languages is an intentional strategy that works within both the hegemonic, neocolonial perspective and the traditional Lakota perspective. Sápa explains that the best way to heal her people from the harms of their colonized history is "a precarious balance of understanding the traditional values and needs of the people as well as being able to work within the governmental system off the reservation to best advocate and work for the people" (Gambrell 2016). In Standing Rock leadership, Dave Archambault II has degrees in business and management, and Chase Iron Eyes is an attorney. Their ability to navigate and work within government, policy, legal, and financial systems is a crucial component of the peaceful, prayerful refusal to accept DAPL's indifference to the potential threat their pipeline poses to the Tribe's water source. For example, in 2016 the water protectors began to call for the banks that finance DAPL to divest. The Defund DAPL website reports over USD $83 million divested by individuals, and over USD $4 billion by tribes, cities, and other organizations (Defund DAPLb). Lakota respect for and facility in the knowledge system of the logic of capital plays out as a strategy of using the tools of the oppressor against the oppressor. Yet if this knowledge of the logic of capital in combination with the Lakota Way of Life can intervene in policy constructively to develop a system of best practices, everyone benefits.

Mastery and Membership

The third factor at issue is how the human place in nature is envisioned in the logic of capital and the Lakota Way of Life. As shown above, the capital logic of domination elevates the human being into the master of nature. In the Lakota Way of Life, humans are simply members of a shared ecosystem. This is evident in the Lakota's conceptions of the human relationship to animals and to water.

Concerning animals, for example, Chase Iron Eyes encounters a group of wild horses while walking the land. He is careful not to get too close as he continues his video post. When the horses become restless at his continuing presence, he leaves in explicit respect for their presence, but also not to cause unpleasantness for himself and the horses by obliging them to confront him. He meets them in a shared place, pays attention, and understands their communication. Subsequently, at a meeting in Italy, he catches on film a local dog who clearly does not want to engage. Iron Eyes says offhandedly, "Maybe I didn't greet him properly in his language." These simple interactions reflect an awareness of and respect for cohabitant species.

In October 2016, a herd of buffalo unexpectedly appeared at Oceti Sakowin. Amid many excited cries, a water protector who by chance was being interviewed at that moment shouts, "Look at all those buffalo! Takanka! Takanka! They are coming for you guys!" (Dewey 2016a). "Takanka" is Lakota for "buffalo," and by "you guys," he clearly means security personnel. From the perspective of capital's logic of domination, he is advocating revenge. Yet he is not angry or frowning; smiling and laughing, he appears joyful at meeting buffalo in this unanticipated but timely way, and he seems to be playfully teasing security personnel who don't know buffalo like he does. They are common in the area (LaCapria 2016), and in Lakota tradition are powerful carriers of messages of greeting, strength, and solidarity from ancestors (Hanson 2016).

Lakota thinking about water comes from a similar but deeper sacredness closely related to place in an experiential cosmology. Tim Mentz, a tribal archive historian and archaeologist, discusses the destruction of Lakota heritage sites in light of the *National Historic Preservation Act* (KOLC-TV 2016). Among other culturally significant items, he identifies several ancestral gathering spots for tribes in a small portion of the DAPL corridor. In particular, he points to two stone rings and three half rings, called arches or crescent moons, located in relation to the handle of Ursa Major, where water would pour onto the land from the stars. This is the spot where *Mni Wiconi* would come to earth and bring a spiritual effigy to life from out of the pouring cup. It is the only place where a chief can fast to show dedication and commitment to leadership. The sites are marked on the ground through collections of small stones overseen by the archaeology firm that came to inspect the area before the USACE granted DAPL the easement. The official archaeologists did not know enough to see that the stones are specifically and precisely placed within the contours of the land and mapped to where the water drains off the great plains. "Every site that we have culturally is linked to water," Mentz says. "*Mni Wiconi*" (KOLC-TV 2016). These sites are where pledges were made and gifts exchanged, always at the meeting point of water – the first medicine. Mentz understands and explains the difficulty academics can experience in trying to reconcile their academic notions of history and archaeology with the nomadic and oral history of the Lakota people in which tradition, ritual, and geography are connected. The sites are not sacred because of what happened there; the meetings and rituals happened there because the places are sacred in the being-in-the-world of water. The events participate in that sacrality, both honouring and being honoured.

These examples demonstrate that the Lakota Way of Life is not a logic of domination. The Lakota ride horses, and traditionally hunt buffalo for food and use their horns and hides during ceremonies. But the animals are not thereby reduced to mere resources. The Lakota are not their master. Water is understood phenomenologically through attention to places that are significant for water's flow. The world is a living network of communications and relationships. The logic of capital and the Lakota Way of Life do not disagree on one point or another; they are different ways of thinking and being in the world.

Authority and Self-Governance

Oceti Sakowin and the Lakota Way of Life are based on shared governance that begins with self-governance rather than the hierarchical authority of top-down command favoured in, for example, corporate, military, and slavery systems. The logic of capital is patriarchal: woman appears as other to man, a dualism that justifies male superiority. For example, the Fort Laramie Treaty could be changed with the agreement of 75 per cent of adult Sioux males. Women did not count for the US treaty writers. Gender in the Lakota Way of Life is embedded in US patriarchy, further complicated by intersectional oppressions and neocolonial impacts on reservation living, and is far too complex to discuss here. Nonetheless, Lakota women's role in initiating the action was significant. The influx of supporters to the Oceti Sakowin camp was originally prompted by Lakota women who organized and posted on Facebook for help stopping the pipeline for the sake of their children (Divided Films 2016). Gambrell (2016) described the women she interviewed (see above) as "getting things done quietly" with focus on *support* for the Tribe and its future, rather than intervening in individuals' choices or actions. This governance through self-selection, for the sake of doing, aimed at community and generational well-being, are paradigmatic of Lakota governance at the Oceti Sakowin camp.

Mainstream reporters covering Standing Rock were unable to identify a leadership structure. A water protector described the camp's governance: "Nobody is really guiding the volunteers to where they are needed, but there's a place for everyone" (Grady 2016). At that time, there were as many as 11,000 people at the Oceti Sakowin camp, making it North Dakota's tenth-largest city (Grady 2016). Archambault described trying to explain to law enforcement officers his lack of authority to order water protectors to take down the camp. They did not answer to him, even though he formally chaired the Oceti Sakowin camp. In December, amid freezing blizzards, when the Army Corps of Engineers recommended against the easement to route the pipeline under Lake Oahe, he thanked supporters and honoured their victory, but said there was no need for anyone to put themselves at further risk (Dewey 2016b). Nonetheless, many remained (Dewey 2016c).

The strength of the movement is its focus on the sacredness of water and life in community within and across generations. The shared goal of water protection is a driving governance principle. Leadership supports and organizes, but hierarchical authority is unnecessary and would be out of place. It would contradict the communal sense of self as being with others in common purpose shared across species and life-sustaining environments. Welcoming activists, military veterans, and buffalo, for example, instantiated being-with in shared experience rather than being-over in a chain of command. The water protectors' clear goals and strategies of peaceful, prayerful actions flow as self-governance from the sacredness of water and future generations, rather than abstract principles of governance and justice. People self-select and cooperate in activities through ground-up governance rather than having their labour managed by the top-down authority of command favoured in corporate, military, and slavery systems, for example. Bottom-up governance is hard to breach – to do so is in effect to turn on oneself. Oceti Sakowin was exceedingly resilient in the face of ongoing assault and extreme weather, not *despite* but precisely *because of* its lack of hierarchical authority. An alleged democracy that mobilizes to support unsustainable industries by assaulting its people when they engage in peaceful protection of resources vital to themselves and future generations has much to learn from their alternative governance model.

Conclusion

We have argued that the water protectors' direct action at the Oceti Sakowin camp demonstrates the incommensurability of the Lakota Way of Life and the logic of capital. The movement was peaceful and prayerful, while capital responded violently with assault and vilification. Water protectors respected the humanity of their assailants and worked within the knowledge system of their oppressor, while DAPL and security forces ignored their voices and were blind to their knowledge system and traditions. While the logic of capital, capable of only short-term thinking, sees man as master of nature and reduces nature to nothing beyond resources to be consumed, water protectors have long-term vision in which people appear as members of a community shared with other people, species, and generations, and in which water plays a special role because of its life-giving and life-sustaining role. Accordingly, water protectors value life over the profit at which the logic of capital aims. Oceti Sakowin accordingly instantiates an alternative logic through its phenomenologic of water and water protectors. In conclusion, we suggest that strengthening the place of this way of thinking in US water policy opens possibilities for healing the destructions of the logic of capital and creating functional policy aimed at protecting people, water, and the environment for the benefit of current and future generations.

Under the Obama administration, in response to the water protectors at Oceti Sakowin, the Department of Justice, the Department of the Army, and the Department of the Interior issued a Joint Statement recommending that the US rethink its relationship to Native American tribes. In particular, the Statement asked for discussion of nationwide reform concerning how tribes' views of projects like DAPL are taken into account, and how to include more meaningful tribal input on decisions affecting their land, resources, and treaty rights, as well as whether new legislation is needed to support these goals of improved tribal stakeholder inclusion. Finally, it asserted the Tribe's right to assembly and free speech, and urged both protesters and those acting for DAPL to act peacefully (Department of Justice 2016). In the spirit of these questions, we make four recommendations concerning US water policy.

First, we recommend that tribal voices be meaningfully included at all levels of water policy decision making, from setting to implementation, in recognition of the Indigenous sovereignty promised in treaties. Sovereignty means Native American lands are within US borders, but are not part of the US. Since water flows across artificially defined borders, shared governance is necessary to protect sovereign rights and tribal well-being. For example, because DAPL does not cross reservation land, Lakota voices were disregarded, despite the fact that the pipeline was routed under their primary water source. Justice demands that sovereign nations be present throughout decision making that has potential impacts on their resources that carry health consequences. Moreover, the necessity for long-term vision in environmentally sustainable decision making that we have discussed above indicates that policy representation of ways of life that have such vision can benefit the whole US population, present and future. We suggest that tribal voices have not just a place at the table, but a role in agenda setting since incommensurability between the logic of capital and an alternative way of life and knowledge system means that the latter perspective has the potential to introduce factors not valued and therefore not considered in the logic of capital.

Second, we recommend reforming the processes for environmental impact assessments (EIA). Given the close relationships between American governance and oil interests, EIAs conducted by an independent third party can be more clearly separated from special interests with oversights in place. Since water does not stay in one place, EIAs should be required regardless of land ownership. EIAs should be performed before, during, and after the project so baselines can be set that contextualize subsequent assessments. Stakeholder assessments identifying groups potentially affected, especially vulnerable groups and marginalized populations, can correlate with EIAs so human rights are protected and costs and benefits of infrastructure development do not breach the requirements of distributive justice. Given Indigenous sovereignty, EIAs relevant to Indigenous nations should be performed in cooperation with tribal representation. Processes should be transparent and EIAs publicly available.

Third, the Department of Justice has an obligation to ensure that credentialled members of the press covering peaceful public expression of their objection to infrastructure projects should retain freedom-of-the-press protections from physical contact, arrest, prosecution, or any other abuse or impediment to their activity. In particular, the arrest of reporters combined with lack of coverage from mainstream news media appears as a strategic limitation of Indigenous voices and voices of dissent. A free press is necessary in democracy, and has a duty to record and report Indigenous actions accurately, so ought to be protected when doing its job.

Finally, we support and recommend greater empowerment and recognition of Indigenous vehicles of governance as sovereign nations in their dealing with the US government. For example, concerning the Black Hills, the court system determines what damages exist and what compensation is due. There is no space for acknowledging what would count as resolution in the Sioux perspective. Such an acknowledgement would support participatory parity (Fraser 1997) in promoting equal participation in decision making, rather than confinement within a system that operates only using a logic of capital so compensation looks like the only resolution. The issue is not whether the Black Hills settlement is enough, but whether it is appropriate when the Sioux actually want their Hills back. Participatory parity is especially important across incommensurable systems, as justice cannot really be experienced by the oppressed on incommensurable terms of the oppressor.

The water protectors of Oceti Sakowin make visible a fractured culture that has lost its connection to life-enabling ecosystems. They disrupt the Eurocentric logic of capital that reduces water to its instrumental value and exhibit a phenomenologic of living-with water in a holistic, community-driven way of life that values people across generations, other species, and ecosystems. If the Lakota and other Indigenous voices not only have a place at the table but also have a voice in setting the agenda, we believe there is hope that the inadequacy of contemporary US water policy to protect water for current and future generations might be remedied, to the benefit of all inhabitants of Turtle Island, human and non-human alike. A phenomenologic of *Mni Wiconi* may in fact be the only hope for teaching global capital that nature does not depend on the economy, but that the economy depends entirely on nature.

REFERENCES

Archambault II, Dave. 2017. "A Violation of Tribal Human Rights: Interview with Amy Goodman." *Democracy Now!* 8 February. Accessed 26 February 2018. https://www.democracynow.org/2017/2/8/a_violation_of_tribal_human_rights/.

Aristotle. 1941. De Partibus Animalium *and* Nicomachean Ethics. *The Basic Works of Aristotle.* Edited by Richard McKeon. New York: Random House.

Atkinson, Scott, and Monica Davey. 2017. "5 Charged with Involuntary Manslaughter in Flint Water Crisis." *The New York Times.* Accessed 26 February 2018. https://www.nytimes.com/2017/06/14/us/flint-water-crisis-manslaughter.html/.

Bacon, Francis. 1980. *The Great Instauration and New Atlantis.* Edited by J. Weinberger. Arlington Heights, IL: Harlan Davidson, Inc.

Bakx, Kyle. 2017. "Keystone XL Could Be Canada's Last Big Oil Export Pipeline." CBS News: Business. Accessed 26 February 2018. http://www.cbc.ca/news/business/kxl-transcanada-oilsands-trump-1.3950256/.

Ball, W.W. Rouse. [1888] 2003. *A Short Account of the History of Mathematics.* New York: Dover Press.

Bibens, Angela. 2016. "11/20 Water Cannon Used on #NoDAPL Protectors." Phone interview by Dallas Goldtooth. SoundCloud. Accessed 26 February 2018. https://soundcloud.com/dallas-goldtooth/1120-water-cannon-used-on-nodapl-protectors-phone-interview-with-angela-bibens/.

Buchele, Mose. 2015. "After HB 40, What's Next for Local Drilling Rules in Texas?" *StateImpact.* Accessed 26 February 2018. https://stateimpact.npr.org/texas/2015/07/02/after-hb-40-whats-next-for-local-drilling-bans-in-texas/.

Businesswire. 2014. "Energy Transfer Announces Crude Oil Pipeline Project Connecting Bakken Supplies to Patoka, Illinois and to Gulf Coast Markets." Accessed 26 February 2018. http://www.businesswire.com/news/home/20140625006184/en/Energy-Transfer-Announces-Crude-Oil-Pipeline-Project/.

Citizens United v. Federal Election Commission. 2010. SCOTUS Blog. Accessed 26 February 2018. http://www.scotusblog.com/case-files/cases/citizens-united-v-federal-election-commission/.

CNN. 2017. "Flint Water Crisis Fast Facts." Accessed 26 February 2018. http://www.cnn.com/2016/03/04/us/flint-water-crisis-fast-facts/index.html/.

Columbia River Inter-Tribal Fish Commission. 2017. Columbia River Treaty. Accessed 26 February 2018. http://www.critfc.org/tribal-treaty-fishing-rights/policy-support/columbia-river-treaty/.

Crazy Bull, Cheryl. 2016. "Woonspe – Education Gives Meaning to Mni Wiconi – Water Is Life." *Indian Country Today.* Accessed 26 February 2018. https://indiancountrymedianetwork.com/education/native-education/woonspeeducation-gives-meaning-to-mni-wiconiwater-is-life/.

Dakwar, Jamil, Director, ACLU Human Rights Program. 2016. "Police at Standing Rock Are Using Life-Threatening Crowd-Control Weapons to Crack Down on Water Protectors." Accessed 26 February 2018. https://www.aclu.org/blog/speak-freely/police-standing-rock-are-using-life-threatening-crowd-control-weapons-crack-down/.

Defund DAPLa. n.d. "Alternatives." Accessed 26 February 2018. http://www.defunddapl.org/copy-of-alternatives/.

– n.d. Accessed 26 February 2018. http://www.defunddapl.org/.

Department of Justice. 2016. "Joint Statement from the Department of Justice, the Department of the Army and the Department of the Interior Regarding Standing Rock Sioux Tribe v. U.S. Army Corps of Engineers." Accessed 26 February 2018. https://www.justice.gov/opa/pr/joint-statement-department -justice-department-army-and-department-interior-regarding-standing/.

Dewey, Myron. 2016a. "The Ancestors Are Always with Us … The Buffalo Remind Us and Validate We Are on the Right Path." Facebook. Posted 10/27/2016.

– 2016b. "Chairman Update #StandingRock." YouTube. Published 12/05/2016. Accessed 26 February 2018. https://www.youtube.com /watch?v=He0M7MkmfQw/.

– 2016c. "'Time to Go Home': Standing Rock Chairman to DAPL Protestors." YouTube. Published 12/06/2016. Accessed 26 February 2018. https:// www.youtube.com/watch?v=aQ1g7ipH8Ao/.

Divided Films. 2016. *Mni Wiconi: The Stand at Standing Rock.* YouTube. Posted 11/14/2016. Accessed 26 February 2018. https://www.youtube.com/watch ?v=4FDuqYld8C8/.

Enbridge. 2017a. "Corporate Social Responsibility." Accessed 26 February 2018. https://www.enbridge.com/about-us/corporate-social-responsibility/

– 2017b. "Who We Are." Accessed 26 February 2018. https://www.enbridge .com/about-us/our-company/.

– 2017c. "Enbridge Inc. Reports Its Fourth Quarter 2016 Results." Accessed 26 February 2018. http://www.enbridge.com/media-center/news/details?id =2124916&lang=en&year=2017/.

Fraser, Nancy. 1997. *Justice Interruptus: Critical Reflections on the "Postsocialist" Condition.* New York: Routledge.

Frommer, Frederic J. 2001. "Black Hills Are Beyond Price to Sioux: Despite Economic Hardship, Tribe Resists US Efforts to Dissolve 1868 Treaty for $570 million." *Los Angeles Times.* Accessed 26 February 2018. http://articles .latimes.com/2001/aug/19/news/mn-35775/.

Fry, Matthew, Adam Briggle, and Jordan Kincaid. 2015. "Fracking and Environmental (In)Justice in a Texas City." *Ecological Economics* 117: 97–107. https:// doi.org/10.1016/j.ecolecon.2015.06.012.

Gambrell, Kem. 2016. "Lakota Women Leaders: Getting Things Done Quietly." *Leadership* 12 no. 3: 293–307. https://doi.org/10.1177/1742715015608234.

Galileo. 1957. *The Discoveries and Opinions of Galileo.* Translated by Stillman Drake. London: Anchor Books.

Gary, Ian, and Terry Lyn Karl. 2003. *Bottom of the Barrel: Africa's Oil Boom and the Poor.* Baltimore: Catholic Relief Services. Accessed 26 February 2018. http:// www.crs.org/our-work-overseas/research-publications/bottom-barrel/.

Gerlak, A.K. 2006. "Federalism and US Water Policy: Lessons for the 21st Century." *Publius* 36 no. 2: 231–57.

Getches, D.H. 2001. "The Metamorphosis of Western Water Policy: Have Federal Laws and Local Decisions Eclipsed the State's Role?" *Stanford Environmental Law Journal* 20 no. 3: 3–72.

Glazebrook, Trish. 2000. "From *physis* to Nature, *technê* to Technology: Heidegger on Aristotle, Galileo and Newton." *The Southern Journal of Philosophy* 38 no. 1: 95–118.

– 2003. "Art or Nature? Aristotle, Restoration Ecology, and Flowforms." *Ethics and the Environment* 8 no. 1: 22–36. https://doi.org/10.2979/ete.2003.8.1.22.

– 2011. "Women and Climate Change: A Case-Study from Northeast Ghana." *Hypatia* 26 no. 4: 762–82. https://doi.org/10.1111/j.1527-2001.2011.01212.x.

Glazebrook, Trish, and Anthony Kola-Olusanya. 2011. "Justice, Conflict, Capital, and Care: Oil in the Niger Delta." *Environmental Ethics* 33 no. 2: 163–84.

Glazebrook, Trish, and Matt Story. 2012. "The Community Obligations of Canadian Oil Companies: A Case Study of Talisman in the Sudan." In *Corporate Social Irresponsibility: A Challenging Concept*, edited by Ralph Tench, William Sun, and Brian Jones.Bingley, UK: Emerald Group Publishing, 231–61.

– 2015. "Heidegger and International Development." In *Heidegger in the Twenty-First Century: Contributions to Phenomenology* 80, edited by Georgakis Tziovanis and Paul J. Ennis. Dordrecht: Springer Science+Business Media, 121–39.

Grady, J. 2016. "DAPL Protest Site Now One of the Largest Cities in North Dakota." YouTube. Published 12/05/2016. Accessed 26 February 2018. https://www.youtube.com/watch?v=Eqkeea3ys8g/.

The Great Peacemakers. n.d. Accessed 26 February 2018. http://www.thegreatpeacemakers.com/iroquois-great-law-of-peace.html/.

Hanson, Hilary. 2016. "Bison Charge Across the Landscape amid Dakota Pipeline Protests." *Huffington Post.* Accessed 26 February 2018. http://www.huffingtonpost.com/entry/bison-buffalo-dakota-pipeline_us_5814d37de4b0390e69d0987c/.

Heidegger, Martin. 1962 [1927]. *Being and Time.* John Macquarrie and Edward Robinson, trans. New York: Harper & Row.

– 1966. *Discourse on Thinking.* Translated by J.M. Anderson and E.H. Freund. New York: Harper & Row.

– 1971. *Poetry, Language, Thought.* Translated by Albert Hofstadter. New York: Harper & Row.

– 1977. *The Question Concerning Technology and Other Essays.* Edited by William Lovitt. New York: Harper & Row.

– 1982. *Nietzsche, Volume 4: Nihilism.* Translated by David Farrell Krell. New York: Harper & Row.

– 1986. *Sein und Zeit.* Tübingen: Max Niemeyer Verlag.

– 1992. *Gelassenheit.* Pfullingen: Günther Neske.

– 1994. *Holzwege, Gesamtausgabe*, Band 5. Frankfurt am Main: Vittorio Klostermann.

- 1995. *Feldweg-Gespräche, Gesamtausgabe*, Band 77. Frankfurt am Main: Vittorio Klostermann.
- 1997a. *Vorträge und Aufsätze*, 8 Auflage. Stuttgart: Neske.
- 1997b. *Nietzsche II, Gesamtausgabe*, Band 6.2. Frankfurt am Main: Vittorio Klostermann.

Heinkel-Wolfe, Peggy. 2014. "Lawsuits Follow Fracking Outcome." *Denton Chronicle*. Accessed 26 February 2018. http://www.dentonrc.com/news /news/2014/11/05/lawsuits-follow-fracking-outcome/.

Herman, Edward, and Noam Chomsky. 2002. *Manufacturing Consent: The Political Economy of the Mass Media*, 2nd ed. New York: Pantheon Books.

Husserl, Edmund. 2001. *The Shorter Logical Investigations*. Translated by J.N. Findlay. Edited by Dermot Moran. New York: Routledge.

Irigaray, Luce. 1985. *Speculum of the Other Woman*. Translated by Gillian C. Gill. Ithaca, NY: Cornell University Press.

Iron Eyes, Chase. 2017. "Water-protectors Call for Mass Mobilizations as Army Plans to Approve Dakota Access Pipeline. Interview with Amy Goodman." *Democracy Now!* 8 February. Accessed 26 February 2018. https://www.democracynow .org/2017/2/8/water_protectors_call_for_global_mass?autostart=true/.

ITA – Information Technology Associates. 2017. "United States Pipelines Map." Accessed 26 February 2018. http://www.theodora.com/pipelines/united _states_pipelines.html/.

Justia: US Supreme Court. 2017. *United States v. Sioux Nation of Indians*, 448 U.S. 371, 376 1980. Accessed 26 February 2018. https://supreme.justia.com /cases/federal/us/448/371/.

Karl, Thomas R., Jerry M. Melillo, and Thomas C. Peterson, eds. 2009. *Global Climate Change Impacts in the United States*. Cambridge, UK: Cambridge University Press.

Kerth, Rob, and Shelley Vinyard. 2012. *Wasting Our Waterways 2012: Toxic Industrial Pollution and the Unfulfilled Promise of the Clean Water Act*. Environment America Research and Policy Center and Frontier Group. Accessed 26 February 2018. http://www.environmentamerica.org/reports/ame /wasting-our-waterways-2012/.

Kneale, Albert H. 1950. *Indian Agent*. Caldwell, ID: Caxton Press.

KOLC-TV. 2016. "Tim Mentz: Updated." YouTube (9:26 mins). Filmed 09/03/2016. Posted 09/17/2016. Accessed 26 February 2018. https:// www.youtube.com/watch?v=w6NapCXUjU0/.

Koerth-Baker, Maggie. 2016. "Police Violence against Native Americans Goes Far Beyond Standing Rock." Accessed 26 February 2018. https:// fivethirtyeight.com/features/police-violence-against-native-americans -goes-far-beyond-standing-rock/.

LaCapria, Kim. 2016. "Standing Rock Protestors Report Wild Buffalo Sighting." *Snopes*. Accessed 26 February 2018. http://www.snopes.com/2016/10/28 /standing-rock-buffalo-sighting/.

Larkin, Molly. 2013. "What Is the 7th Generation Principle and Why Do You Need to Know about It? Ancient Wisdom for Balanced Living." Accessed 26 February 2018. https://www.mollylarkin.com/what-is-the-7th-generation -principle-and-why-do-you-need-to-know-about-it-3/.

Lemmen, Donald S., and Fiona J. Warren, eds. 2004. *Climate Change Impacts and Adaptations: A Canadian Perspective*. Ottawa, ON: Natural Resources Canada. Accessed 26 February 2018. https://cfs.nrcan.gc.ca/publications /download-pdf/27428/.

Levin, Sam. 2017. "Army Veterans Return to Standing Rock to Form a Human Shield against Police." Accessed 26 February 2018. https://www.theguardian .com/us-news/2017/feb/11/standing-rock-army-veterans-camp/.

List of oil spills. 2017. *Wikipedia*. Accessed 26 February 2018. https:// en.wikipedia.org/wiki/List_of_oil_spills/.

List of pipeline accidents in the United States in the 21st century. 2017. *Wikipedia*. Accessed 26 February 2018. https://en.wikipedia.org/wiki /List_of_pipeline_accidents_in_the_United_States_in_the_21st_century/.

Lott, Fraser C., Nikolaos Christidis, and Peter A. Stott. 2013. "Can the 2011 East African Drought be Attributed to Human-induced Climate Change?" *Geophysical Research Letters* 40 no. 6: 1177–81. https://doi.org/10.1002 /grl.50235.

May, Charlie. 2016. "A Reporter's Notebook: The Journey to Oceti Sakowin, the Protest Camp of the Standing Rock Sioux." Accessed 26 February 2018. http://www.salon.com/2016/12/25/a-reporters-notebook-the-journey-to -oceti-sakowin-the-protest-camp-of-the-standing-rock-sioux/.

Mother Jones and the Foundation for National Progress. 2017. "A History of Native Americans Protesting the Dakota Access Pipeline." *WorldPress.com* Accessed 26 February 2018. http://www.motherjones.com/environment /2016/09/dakota-access-pipeline-protest-timeline-sioux-standing-rock -jill-stein/.

Moyers, Bill. 2016. "What You Need to Know about the Dakota Access Pipeline Protest." *Common Dreams*. Accessed 26 February 2018. https:// www.commondreams.org/views/2016/09/09/what-you-need-know-about -dakota-access-pipeline-protest/.

National Transportation Safety Board. 2010. "Crude Oil Pipeline Rupture and Spill." Accessed 26 February 2018. https://web.archive.org/web /20120630002212/http://www.ntsb.gov/investigations/2010/marshall _mi.html/.

– 2012. "Enbridge, Inc. Hazardous Liquid Pipeline Rupture." Accessed 26 February 2018. https://www.ntsb.gov/investigations/AccidentReports /Reports/PAR1201.pdf/.

Neihardt, John G. 2008. *Black Elk Speaks: Being the Life Story of a Holy Man of the Oglala Sioux*. Albany, NY: State University of New York Press.

Nicholson, Blake. 2017. "Dakota Access Protest Policing Costs exceed $22M." *The Seattle Times.* Accessed 26 February 2018. http://www.seattletimes.com /business/dakota-access-protest-policing-costs-exceed-22m/.

Petersen, Tom. 2016. "Why I Joined My Fellow Vets at Standing Rock This Weekend." Accessed 26 February 2018. https://www.aclu.org/blog /speak-freely/why-i-joined-my-fellow-vets-standing-rock-weekend/.

Pipeline 101. 2016. Accessed 26 February 2018. http://www.pipeline101.org/.

Samson, Andrew. 2008. *Water in Texas: An Introduction.* San Marcos, TX: Texas State University Press.

Scahill, Jeremy. 2010. "Blackwater in Pakistan: Gates Confirms." *The Nation.* Accessed 26 February 2018. https://www.thenation.com/article /blackwater-pakistan-gates-confirms/.

Shiva, Vandana. 1991. *The Violence of the Green Revolution,* London: Zed Books.

Standing Rock Sioux Tribe. 2014. "Sep 30th DAPL Meeting with SRST." (1:08:17) Recorded 09/30/2014. Posted 12/03/2016. Accessed 26 February 2018. https://www.youtube.com/watch?v=ZlwdtnZXmtY/.

Stevens, Craig. 2016. "On the Dakota Access Pipeline, Let's Stick to the Facts." *The Hill.* Accessed 26 February 2018. http://thehill.com/blogs/congress -blog/energy-environment/296926-on-the-dakota-access-pipeline-lets-stick -to-the-facts/.

Streshinsky, Maria. 2011. "Saying No to $1 billion: Why the Sioux Won't Take Federal Money." *The Atlantic.* Accessed 26 February 2018. https://www .theatlantic.com/magazine/archive/2011/03/saying-no-to-1-billion /308380/.

Tannous, Nadya Raja. 2016. "Palestinians Join Standing Rock Sioux to Protest Dakota Access Pipeline." *Mondoweiss.* Accessed 26 February 2018. http:// mondoweiss.net/2016/10/palestinians-standing-pipeline/.

Three Legs, Bobbi Jean. 2017. "Water Is Life. Water Is Sacred." *Democracy Now!* Accessed 26 February 2018. https://www.democracynow.org/2017/1/25 /water_is_life_water_is_sacred/.

Vecsey C., and R.W. Venables. 1994. *An Iroquois Perspective: American Indian Environments: Ecological Issues in Native American History.* Syracuse, NY: Syracuse University Press.

Vogel, Kenneth P. and Eliana Johnson. 2016. "Trump's Koch Administration." *Politico.* Accessed 26 February 2018. http://www.politico.com/story/2016 /11/trump-koch-brothers-231863/.

Warren, Karen. 1996. *Ecological Feminist Philosophies,* ed. Karen Warren. Bloomington, IN: Indiana University Press.

11 Phenomenology, Water Policy, and the Conception of the *Polis*

HENRY DICKS

From Phenomenology to Water Policy

What is the relevance of phenomenology to water policy? Phenomenology is the attempt to uncover the hidden and primary features of worldly experience, which, it is assumed, have become obscured by various acquired prejudices. This method may be applied to our lived relation to water. What is our primary way of experiencing water? The answer to this question may in turn provide the basis for water policy, inasmuch as it allows us to concentrate on what is most primary and essential in our relationship to water, while at the same time articulating this primary and essential relation to water with other, secondary ways of relating to water. In this first section, we will consider briefly how three phenomenologists – Husserl, Heidegger, and Levinas – have each sought to understand the primary features of worldly experience, before going on to consider how these may be: 1) applied to water; and 2) translated into water policy.

For the later Husserl of the *Crisis of European Philosophy and Transcendental Phenomenology* (1970), the scientific way of seeing, often presumed to provide our only access to the real, is secondary to our lived experience of things in what he calls the "life-world," and it is thus necessary to bracket the scientific way of seeing, a procedure referred to as "*epoché*." Husserl (1981) elsewhere provides a famous example of the secondary nature of scientific representation: our relation to the earth. In treating the earth *as a body*, science overlooks what Husserl argues is the primary relation to the earth – the earth as *ground* (see Himanka 2005). But prejudices have often been detected not just outside of phenomenology, but also in the work of other phenomenologists, despite their claims to access in a direct and unmediated manner the fundamental traits of human experience. Thus it was that in *Being and Time*, Heidegger (1995) argued that the Western tradition from Descartes to Husserl[1] had assumed that the "things themselves" are manifest primarily in the mode of what he calls the "present-at-hand." According to Heidegger, however, things appear in the first instance not as "present-at-hand" objects, but rather as "ready-to-hand" tools, understood not as objects possessing observable properties (colour, size, weight, etc.), but rather as equipment immediately useful for accomplishing some project or other. Levinas (1961), in turn, argued that Heidegger's own phenomenological account of being-in-the-world was in fact secondary to a more primary mode of experience. For Levinas, experience is not in the first instance of "things" at all, but rather of "qualities without substances." Primordial experience, on this view, is not of observable objects or of usable pieces of equipment, but rather of bathing in and assimilating what he calls the "elemental." In a comparable vein, Bruce Foltz has argued that the later Heidegger no longer saw our primary experience of things as ready-to-hand entities in a functional network, but rather as a "basically poetic encounter with primordial nature" (Foltz 1995, 51).

These phenomenological attempts to uncover a primary layer of experience do not imply that, once uncovered, the primary layer will become the only one. Husserl did not deny the scientific truth that the earth qua body revolves around the sun, but argued, rather, that this view has problematically obscured our primordial relation to the earth qua ground. Heidegger did not deny that things manifest themselves to us as present-at-hand objects, but rather that this mode of appearance only occurs when their tool function breaks down. Levinas did not deny that we have access to "things," but he saw the experience of things as a secondary phenomenon – unique to humans and arising ultimately from a need to possess and store discrete entities for future consumption – imposed upon a more fundamental experience, common to life in general, of bathing in elemental qualities. And the later Heidegger did not abandon the idea that in our average, everyday existence we relate to things as ready-to-hand tools, but sought rather to show that it is in great poetry and philosophy that the different historical worlds we inhabit are brought forth, thus providing the basic framework in which our relations to things and our relation to other humans may take place. In every case, then, the various different layers of existence are not only analysed in terms of their relative primacy, but also articulated in such a way that the relations and transitions between these layers may also become apparent.

It is not difficult to see that each of the phenomenological analyses of existence outlined above could be applied to water. For the later Husserl, the bracketing of the scientific way of seeing would mean that water would not, in the first instance, be what it is in its scientific description (a chemical compound composed of hydrogen and oxygen, a fluid obeying the laws of hydrodynamics, etc.), but rather an entity we encounter in the life-world in the form of rivers, lakes, rain, and so on. For the later Heidegger, our primary relation to water would be not to an object of scientific or everyday knowledge, and not even to "something-in-order-to" (*etwas-um-zu*) – always already in use for irrigation, transportation, cooking, and so on – but rather to a part of the world whose place and role is set forth in great poetry (and philosophy). Indeed, it is often precisely with respect to water, and rivers more specifically (the Rhine, the Ister), that Heidegger develops his reflections on the primacy of the poetic (Heidegger 1980; 1996). Lastly, for Levinas, our primary relation to water would not be to water as a "thing" at all, but rather to an elemental quality in which we joyfully "bathe."

Each of these phenomenological approaches to water could in turn be translated into policy. In reversing the traditional view that scientific descriptions of water give access to what is primordially real, whereas everyday experiences of water are mere subjective representations, Husserlian phenomenology could lead to the development of forms of water policy that accord primary importance to everyday experiences and representations of water manifest in the life-world,[2] rather than to policies based primarily on scientific expertise and which run the risk of blaming concrete water-related problems on the inability of everyday water users to align their representations and correlative behaviour with those of scientifically informed experts.

A Heidegger-inspired approach, by contrast, could involve studies of how people actually interact with water in pragmatic contexts – in the home, when watering the lawn, when sheltering from the rain, and so on – and then look to see how these always arise within the context of a historically specific configuration of things and persons as set forth in great works of poetry (and philosophy). From this perspective, one could further argue that, faced with a generalized breakdown in our functional systems (including our water systems) arising as a result of such contemporary problems as climate change, resource depletion, widespread pollution, falling biodiversity, and so on, what is required is not simply an attempt to study and repair the broken elements of the current system (via recourse to their study qua present-at-hand objects), but rather, and as the precondition of a transition to another system, the elaboration of great works of poetry and philosophy that instantiate or present new ontologies capable of bringing forth new configurations of things and people, and therewith also what Heidegger calls a "new beginning." Lastly, a Levinassian approach would no doubt be critical of an instrumental approach to water, and would instead underline the importance of two basic phenomena: enjoyment and possession. Translated into water policy, this could lead to policies that emphasize and celebrate qualitative experiences of water, such as swimming in the sea, hot baths, or drinking cool mineral water, while also recognizing that, in the case of humans, access to these enjoyable, qualitative experiences is dependent on the reification, quantification, and possession of water, which in turn implies inevitable economic and environmental inequalities. Not everyone has access to clean drinking water, adequate sanitation, unpolluted rivers and beaches, or spa treatments and private swimming pools, and many are not only deprived of the qualitative enjoyment of water these amenities afford, but are also exposed to unpleasant experiences of untreated sewage and industrial wastes, the loss of loved ones to water-borne diseases, catastrophic droughts and flooding caused partly by the carbon emissions of others more fortunate, and other water-related causes of suffering. Drawing on Levinas's concepts of the infinitely Other (*l'infiniment Autre*) and the Third (*le tiers*), these inequalities could in turn be combatted both by altruistic acts of charity and, as far as water *policy* is concerned, by political attempts to distribute "aquatic goods" (clean drinking water, access to surface waters, etc.) more equitably, while also eliminating, where possible, the "aquatic bads" (pollution, water-borne disease, etc.) to which the least well-off are disproportionately exposed.

This is not to say, of course, that the transition from phenomenology to water policy is straightforward in the sense of being *directly deducible* from the basic phenomenological stance one adopts, for hermeneutics will also inevitably play a role. If one starts with the work of a given phenomenologist, then one clearly needs to *interpret* their work, to *apply* it to the specific case of water, and thereafter to *translate* this applied phenomenology into water policy – three processes that call for much in the way of judgment and thus constitute potential sources of disagreement. Nevertheless, it is also true that *something like* the transitions from phenomenology to water policy I have sketched above are manifestly possible, in which case there can be little doubt that phenomenology can provide a powerful source for both framing and developing different approaches to water policy.

Phenomenology and Ontology: From the Things Themselves to Being Itself

It seems fair to say that the brief sketches presented above of how the phenomenology of Husserl, Heidegger, and Levinas might be translated into water policy are all quite plausible. This plausibility is reinforced when one considers that each of these philosophers may be seen as offering a phenomenological approach to one or more long-standing fields of human activity: Husserl's phenomenology relates primarily to knowing, and therewith also to science and epistemology; Heidegger's to using and making, and therewith also to technics and poetics; and Levinas's to enjoying, owning, giving, and sharing, and therewith also to economics, ethics, and politics.

In view of this, it could perhaps be argued that what is ultimately required is an approach to water policy that does not seek to choose between Husserl, Heidegger, and Levinas, as if only one of epistemology, technics, and ethics were what really mattered, but rather a more comprehensive water policy based on the articulation of phenomenological approaches to these three spheres of human experience, and thus capable of accounting for a wide variety of different ways of understanding and interacting with water. After all, do we really have to choose between seeing water as an object of scientific/everyday knowledge (Husserl), as a practical instrument whose role in the world derives ultimately from our relation to Being as set forth in poetry and philosophy (Heidegger), or as an elemental quality amenable to reification, quantification, and possession by humans (Levinas)? Is it not rather the case that all of these different modalities of the "as" are valid in their own spheres, thus requiring careful articulation, rather than being placed in competition for existential primacy? Just as each of the three phenomenologists we have considered articulates a limited number of layers of experience or ways of relating to things, a more comprehensive phenomenological framework would articulate the work of all these different phenomenologists (and perhaps others as well), thus opening up the possibility of their conjoined translation into a more comprehensive approach to water policy.

This in turn raises the question of how one might go about this articulation. With a view to answering this question, let us consider an issue of fundamental importance to phenomenology that we have not yet directly considered: the question of Being. Ever since Husserl, a basic claim of all phenomenology has been that our experience of things always involves seeing things "as" something or other; to see is to see "as." In keeping with this, in our earlier discussion of various ways of seeing water, it was in every case *as* something or other that water appeared: as a correlate of subjective consciousness, whether scientific or prescientific (Husserl); as a ready-to-hand tool in a functional network (Heidegger); or as an elemental quality amenable to reification, quantification, and possession (Levinas). This in turn points to the possibility not of identifying one mode of Being as primary, one modality of the "as" as more fundamental than the others, but rather of thinking about the "as" itself, as opposed to the "as something." With this in mind, note that there is a fundamental difference between Heidegger's approach to phenomenology and the approaches of Husserl and Levinas. In my discussion of Heidegger's approach above, I wrote that it entails the view of water as a "practical instrument whose role in the world derives ultimately from our relation to Being as set forth in poetry and philosophy." What I did not point out was that this view clearly articulates two levels of analysis: on the one hand, there is the view of water as a practical instrument; on the other, there is the idea that how we relate to water as a practical instrument derives from a deeper source, from our relation to Being, as set forth in poetry and philosophy. This suggests that if we are to articulate the work of Husserl, Heidegger, and Levinas, it is perhaps not so much on the basis of an attempt to hierarchize their phenomenological approaches to technics, ethics, and epistemology, but rather on the basis of a new way of thinking about Being itself, a new ontology, which may be expressed either implicitly in poetry or explicitly in philosophy.

The Opening of Being: A Two-Stage Process

Heidegger, it is well known, did not simply reflect on Being, on the "as," but also on the opening or clearing in which Being appeared. This opening or clearing, he further thought, is unique to humans and constitutes the primordial site in which "things" come to presence (Heidegger 1993). Whereas animals experience only environmental "triggers," which have a disinhibiting effect on their behaviour (Heidegger 1995), humans have access to "beings," understood as distinct entities that may be understood "as" something or other, a way of understanding Being that Heidegger calls "pre-ontological," in contradistinction to the explicit theorization of Being characteristic of ontology.

One important but largely ignored complication with distinguishing between humans and animals in this binary manner is that evolutionary theorists often conceptualize the transition from animality to humanity not as a one-step process but as a two-step one. The first of these steps corresponds to the emergence of the genus *homo*, or "early humans," as they are often called. Quite when this happened and which was the first true species of *homo* is a disputed matter. The French paleoanthropologist Pascal Picq (2009), for example, thinks that *homo habilis* and *homo rudolfensis* were not true members of the genus *homo*; instead, he attributes this status only to *homo ergaster* (or perhaps *homo erectus*) and its descendants. But wherever one situates the transition, the idea that there occurred an important rupture with our primate ancestors around

1.5 to 2 million years ago and that this marked the birth of the genus *homo* seems to be largely accepted among the scientific community. Important changes associated with this step from animality to early humanity include a rapid increase in brain size, systematic recourse to tool use, and modifications in social relations, all of which, it is often considered, evolved in response to the transition from predominantly forested environments to predominantly open ones (Hublin 2008).

The second step occurred between 200,000 and 50,000 years ago with the birth of *homo sapiens* or, more colloquially, "modern humans." One hypothesis that has been put forward to explain this second transition is that of the British cognitive archaeologist Steven Mithen (1996). Drawing on both archaeological evidence and research into evolutionary psychology, cognitive science, and philosophy of mind, Mithen argues that the distinguishing feature of *homo sapiens* is what he calls "cognitive fluidity." According to Mithen, pre-*sapiens* members of the genus *homo* possessed three main domains of intelligence: 1) natural history intelligence, which is required to navigate a landscape, interact in various ways with different species, and so on; 2) technical intelligence, which centres on the use of the hands to fashion and manipulate tools; and 3) social intelligence, which involves a recognition of the Other as Other and therewith also a potential attentiveness to their specific needs and goals.[3] For Mithen, what sets *homo sapiens* apart from other species of *homo* is not so much the emergence of a new domain of intelligence, but rather the ability to combine these pre-existing spheres of intelligence fluidly. Mithen provides numerous examples of this "cognitive fluidity," including the use of tools for social purposes, as is the case in bodily decoration (technical meets social intelligence), the fabrication of tools out of bone and other materials sourced from animals (technical meets natural history intelligence), and the prevalence of totemism and animism in contemporary hunter-gatherer societies (natural history meets social intelligence).

But what, one might wonder, has all this got to do with phenomenology and the question of Being? Husserl and Heidegger went to great trouble to distinguish their phenomenological methods from those of the positive sciences, with positive science typically being discussed only in negative terms, as something secondary to phenomenology, as in Husserl's discussion of Copernican physics.[4] In recent times, the strategy common to both Husserl and Heidegger of opposing phenomenology to the positive sciences, and then concentrating exclusively on the former, has been called into question by researchers seeking to articulate constructively their different forms of insight.[5] A particularly interesting example of this for our purposes is Peter Sloterdijk's attempt to shed light on Heidegger's concept of the "clearing" through the study of human evolution (Sloterdijk 2017). Without going into the details of Sloterdijk's argument, what matters in the present context is that his method of articulating paleoanthropology with Heideggerian thinking about Being (and the clearing), which he playfully calls "paleo-ontology," may be applied to the two-step transition between animality and humanity. Applying Sloterdijk's method in this manner suggests an interesting possibility: that much of what phenomenologists typically attribute to "humans" may already have been in place in so-called early humans (especially *homo erectus*). The rapid development of technical intelligence that occurred with the birth of early humans, for example, suggests that the average, everyday understanding of beings as in the first instance "something-in-order-to" may already have been in place 1.5 million years ago. Likewise, the increase in social intelligence – including apparent cases of altruistic caring for infirm individuals (Mithen 2006, 135) – suggests that what Heidegger calls "being-with" may also have characterized the being-in-the-world of early humans. If this is the case, then it would seem that early humans may have already left behind the state of animality – understood as the possession of an encircling ring amenable to disinhibition by environmental triggers – and acquired a "pre-ontological understanding of Being," that is to say, an understanding of beings "as" this or that, and in particular as "something-in-order-to" (in the case of non-humans) or as "Others" with whom we are always already in the world (in the case of humans). Arguments of this sort are of course highly speculative. It involves significant speculation to argue from archaeological evidence to concrete positions in evolutionary psychology and anthropology, and the transition from the science of human evolution to "paleo-ontology" is even more speculative. But this is not to say that "paleo-ontological" speculation should be prohibited a priori, as if human science alone were the sole possible mode of analysis of our distant ancestors, only that the sort of "fantastic reconstructions" made possible by "paleo-ontology" will inevitably be judged by other standards than those of either science or philosophy *on their own*, including overall plausibility, cogency, fertility, explicative power, and so on.

If one accepts, at least as a working hypothesis, the idea that the preontological understanding of Being may already have been present in early humans, this then raises the question of how the relation to Being changed with the emergence of modern humans. The answer, I will now suggest, is that, with cognitive fluidity, the "as" attained a state of "openness," not in the sense that it could be understood pre-ontologically, for in our hypothesis that was already the case with early humans, but rather in the sense that it henceforth became possible *to see one being "as" another being*, in pre-ontological understanding of the form "A as B," A and B could henceforth stand for anything at all, at least in principle. Being, on this view, would already have been open to the pre-ontological understanding of early humans, but they would have been incapable of fluidly blending together the "Being" of different entities, thus also imagining and creating new hybrid forms of existence. With early humans, an animal could have been a source of food or a danger, but only with modern humans could it have been an ancestor, a god, a prophetic sign, a totem, and so on (Stringer and McKie 1996, 212). Moreover, I would argue that it was with the advent of cognitive fluidity that poetry became possible, where poetry is understood, in a sense close to myth, as narrations of the coming into being and unfolding of the world, something that typically depends on cognitively fluid crossings over and transitions between gods, men, animals, plants, tools, the elements, and so on.

We are now in a better position to give a more precise meaning to our Heidegger-inspired claim that Being is set forth above all in poetry (and later, also philosophy), and that it is this setting forth that fundamentally determines the basic configuration of the different entities – natural, technical, human, etc. – that make up the world. The basic configuration of the world is a result of the "openness" of Being in the sense outlined above, and it is within the configurations produced by this openness that different beings find their meaning and their place. This in turn allows us to see the limitations of a phenomenological approach to water derived from Husserl, early Heidegger, or Levinas (at least as their thought was interpreted and presented above). While it is true that water variously manifests itself as an object of everyday and scientific knowledge, as a tool in a network of functional relations, and as both an elemental quality and a quantifiable thing to be possessed, given, and shared, it is also true that the basic configuration of the world in which all of these different modes of existence are concretely instantiated and co-articulated is the result of a deeper poetic conception and articulation of the basic elements of worldly existence.

Two Conceptions of the Polis: The Polis as Human Organism and Forest Ecosystem

One obvious objection that could be made to the idea that the place of water in the world derives ultimately from some kind of deep poetic conception and articulation of the different domains of Being is that, with the advent of reason, clear and distinct ideas, positive science, and so on, we have been able to overcome the primitive confusions produced by cognitive fluidity. No longer do we mingle together and confuse nature, tools, and people, let alone tell tales of their interactions with gods and other mythical entities; on the contrary, we separate them out into different disciplines studied respectively by natural scientists, engineers (and designers), and social scientists. In what follows, I will respond to this objection by showing how, despite frequent claims to have left primitive confusions of the different realms of Being behind, the basic configuration of the world has always depended fundamentally on cognitive fluidity. More specifically, I will argue that what ultimately determines the instantiation and articulation of technics, ethics, and epistemology is the way that humans *conceive* their collective home, where the word "conceive" is to be understood in both senses of the word: as both "understanding" and "designing." From this perspective, the ultimate place of water in the world is, at least in the Western tradition, a consequence of how we conceive the polis, the original collective home of the people of the West.

Ever since Plato, it has been common to conceive the polis by analogy with a human being. Plato himself conceived the polis by analogy with what he saw as the three parts of the human soul (Desmond 1974). John of Salisbury (1990) conceived the polis by analogy with the human mind and body, a way of thinking that was later taken up by Hobbes, who in *Leviathan* (1651) conceived the city as an "artificial Man," as well as by other contract theorists, particularly Rousseau (2007). Similarly, in the early Italian Renaissance, such important architects and urban theorists as Alberti, Filarete, and Giorgio Martini all conceived the city as having the basic form of the human body (Choay 1974). This conception of the polis continued throughout the nineteenth century, though it also underwent something of an inflection. In the field of urbanism, the city was no longer conceived as a body, but rather as an organism, hence the extensive analysis advanced by Cerda of the city as an "urban organism" (Choay 1996). At the same time, as society emerged as a separate category from the State, the nascent science of sociology conceived human society physiologically as an organism, as was the case most notably in the work of Saint-Simon (1965), Comte (1995), and Durkheim (2011).

This conception of the polis as a (human) organism had a major influence on the role accorded to water. An important example of this is the work of the British hygienist Frederick Ward (1856), who argued that the circulation of water in the city should follow the basic model of the circulation of blood in the human organism. Just as water circulates in the human organism along veins, arteries, and capillaries via the pumping action of the heart, Ward thought it should circulate in the city and in society in much the same way. An important consequence, as Ivan Illich puts it in "H$_2$O and the Waters of Forgetfulness," was that "[l]ike the individual human body and the social body, the city was now also described as a network of pipes" (1986, 45). Similarly, according to the French urban hydrologist Bernard Chocat (2015), the generalization of impermeable paving in the city, which also began in earnest in the nineteenth century, was based on an analogy with the skin. Like the skin, the impermeable surfaces of the city would have two functions: the physical one of preventing water from penetrating the urban organism, and the immunological one of protecting the urban organism from the threat of water-borne diseases. Further, this conception of the city, applied in practice primarily by engineers, architects, and urbanists, often under the guiding influence of physicians, for whom the model of the human body came quite naturally, was also visible in the literature of the time. Victor Hugo wrote in *Les Misérables* (1862) of the plight of sewage workers forced to descend into the "intestines of Leviathan," and Émile Zola's *Le ventre de Paris* (1878) conceived the central market of Les Halles as the "belly" (*ventre*) of the city. In doing so, however, they were not simply speaking metaphorically, but, in the case of Zola, echoing the way the city had been conceived by Renaissance architects and urbanists, who had argued that the marketplace should be situated at the city centre, in a position corresponding to the belly button of a human lying on their back, with legs and arms outstretched, as in the famous image of "Vitruvian man" (Choay 1974), and, in the case of Hugo, repeating the founding analogy of nineteenth-century engineers, who conceived the circulation of water in the city by analogy with its circulation in the human body.

If the conception of the city as a human organism, replete with analogues to the heart, veins, arteries, capillaries, skin, stomach, bowels, and so on, has indeed equipped the city with a functioning system for water management, it is not without problems (Dicks 2015a). As cities conceived according to the model of the human body or organism expand, they demand ever greater quantities of clean water, often sourced from distant sources or fossil reservoirs, while at the same time requiring ever more extensive (and expensive) sewage evacuation and treatment systems. Further, as native ecosystems are smothered by impermeable paving, urban biodiversity declines, the urban heat island effect becomes a problem, flash floods become an increased danger, and stormwater treatment costs also increase (whether in unitary or combined systems).

Faced with problems such as these, an increasing number of architects and urbanists, especially those working in the field of biomimicry, have proposed basing cities on the model of native ecosystems, in particular the forest (Schuiten 2010; Callebaut 2015). "Imagine a building like a tree, a city like a forest," write Braungart and McDonough (2009). This involves thinking about the place of water in the city in a radically different way (Dicks 2015b). Just as in forests water often flows in rivers and streams, so it would be possible for those rivers and streams that have been covered over, polluted, and integrated into the "intestines of Leviathan" to be restored to the light of day. Just as in forests permeable soils allow water to infiltrate the soil directly, thus regulating water flows in a way that reduces the danger of both local and downstream flooding, while also ensuring the presence of water for future use, so it is possible to employ various techniques – permeable paving, vegetal surfaces, drainage swales, etc. – to allow urban water to infiltrate the soil directly, where it may function as an additional resource. Just as in forests the primary source of water used by the trees and other plants is local rainfall, so it would be possible to design the built environment to facilitate local rainwater collection and retention, thus reducing dependence on long-distance transfers and fossil reservoirs. Just as in forests various physical and biological filtration processes ensure that water is cleaned of impurities, so similar techniques – phytodepuration, mycofiltration – may be deployed in cities. Just as water passing out of forest ecosystems will often flow into wetlands, where it will undergo further purification, so wastewater passing out of the city could be treated in modified natural wetlands, constructed wetlands, or "living machines" modelled on natural wetlands (Todd and Todd 1993). And just as in forests evapotranspiration constitutes an effective cooling mechanism for the ecosystem as a whole, so buildings could be kitted out with functional equivalents to the forest canopy, such as green roofs and façades, capable of cooling the city in much the same way.

Taking a step back from these practical applications, it is clear that the cognitively fluid way of thinking that underlies them is compatible with our Heidegger-inspired analysis of the openness of Being, for it is this openness that makes possible differing conceptions of the polis, which in turn give rise to different ways of using water characteristic of differently configured urban water systems. Nevertheless, while we have focused in the first instance on the broadly Heideggerian issue of the role of water in differently organized functional systems, it is also true that these differing organizations will have a strong influence on how water is enjoyed, owned, and shared among the population (Levinas), as well as on the sorts of everyday and scientific knowledge produced in relation to it (Husserl). So, while the approach to water policy outlined above does seem to suggest a certain priority of technics and poetics over a Levinassian focus on ethics and politics, or a Husserlian one on science and epistemology, it also opens up the possibility that all three of these phenomenological approaches to water could be analysed in relation to the underlying "conception" of the polis on which they depend for their concrete realization.

Conclusion

Returning to the issue of phenomenology's relation to water policy, the principal message of the above analysis is that the fundamental issue that needs to be addressed in determining water *policy* is how we conceive the polis (and the place of water therein), for it is our conception of the polis that ultimately provides the basic framework within which specific water policies may take shape. This in turn implies that obvious applications of phenomenology to water policy – based on water's manifestation as a tool in a functional network (early Heidegger), as an elemental quality and quantifiable possession (Levinas), or as an object of everyday or scientific knowledge (Husserl) – are radically inadequate, though this is not to say that each approach is not valid in its own sphere. What is lacking in each of these cases is an understanding of the fact that the fundamental configuration of the world results ultimately from the openness of Being, understood as the capacity of the "as" to link beings from radically different spheres of existence, thus making it possible to cross over in myriad ways the various domains of thought, a faculty that Mithen has called "cognitive fluidity." As far as water is concerned, the most important manifestation of cognitive fluidity concerns how we conceive the polis, qua both city and State. Depending on whether we conceive the polis by analogy with the human body or organism, as has traditionally been the case, or by analogy with a forest ecosystem, as is increasingly being advocated in contemporary urban theory, water will manifest itself in fundamentally different ways. So, while in both conceptions of the polis, water will of course appear as a functional element embedded in multiple ways in a wider system, as an elemental quality and quantifiable possession, and as an object of everyday and scientific knowledge, in each case the underlying framework within which these manifestations of water are embedded will be radically different. It is this underlying framework, this fundamental conception of the polis, that is ultimately of primary importance to water policy.

ACKNOWLEDGMENTS

I would like to thank the environmental services firm Suez and the LabEx IMU (Intelligences des Mondes Urbains) of the University of Lyon for the financing of two postdoctoral research positions – the first on water, the second on biomimetic cities – during which many of the ideas presented in this chapter were first developed.

NOTES

1 Husserl was not a direct object of Heidegger's critique in this text, but there can be little doubt that the phenomenology Husserl had developed before *Being and Time* belongs to the tradition Heidegger was attempting to deconstruct. Moreover, it is also important to realize that Husserl's concept of the "life-world," put forward several years after *Being and Time*, is often considered to have been developed at least in part as a response to Heidegger's analysis of "being-in-the-world."

2 For Husserl, these experiences would of course need to be subjected to rigorous philosophical analysis via a second, "transcendental" *epoché* that would bracket out what he calls the "natural attitude" and instead seek to analyze experience in terms of the relation between the transcendental ego and the world as correlate of intentional consciousness.

3 Mithen also talks about linguistic intelligence, but it would be beyond the scope of this brief exposition of his thought to discuss here his view of its complex relation to the other domains of intelligence.

4 There are some interesting exceptions to this, notably Heidegger's discussion of various positive experiments in biology in *The Fundamental Concepts of Metaphysics* (1995).

5 Much of this work has taken place at the interface with cognitive science, as is the case in the "neurophenomenology" of Varela (1996).

REFERENCES

Braungart, Michael, and William McDonough. 2009. *Cradle to Cradle: Re-Making the Way We Make Things*. London: Vintage.

Callebaut, Vincent. 2015. *Paris 2050: Les cités fertiles faces aux enjeux du XX siècle*. Paris: Michel Lafon.

Choay, Françoise. 1974. "La ville et le domaine bâti comme corps dans les textes des architectes-théoriciens de la première renaissance italienne." *Nouvelle Revue de Psychanalyse* 9: 239–51.

– 1996. *La règle et le modèle: Sur la théorie de l'architecture et de l'urbanisme*. Paris: Éditions du Seuil.

Chocat, Bernard. 2015. "L'invention des eaux usées au XIXe siècle et ses conséquences sur la gestion des eaux urbaines d'aujourd'hui." In *L'usé, le sale, l'impur: Rationalités, usages et imaginaires de l'eau*, edited by Cécile Nou, Claire Harpet, Jean-Philippe Pierron, and Henry Dicks. 37–46. Louain-la-Neuve: EME.

Comte, Auguste. 1995. *Leçons de Sociologie: Cours de philosophie positive*. Edited by Juliette Grange, 47–51. Paris: Flammarion.

Desmond, Lee, trans. 1974. *Plato's Republic*. London: Penguin Classics.

Dicks, Henry. 2015a. "De la ville anthropomimétique à la ville biomimétique: les eaux usées, sales et impures dans le nouvel imaginaire de la ville forêt." In *L'usé, le sale, l'impur: Rationalités, usages et imaginaires de l'eau*, edited by Cécile Nou, Claire Harpet, Jean-Philippe Pierron, and Henry Dicks. 37–46, Louain-la-Neuve: EME.

– 2015b. "Penser le nouveau paradigme de l'hydrologie urbaine: bio-mimétisme, éco-phénoménologie, et gestion intégrée." *La Houille Blanche* 5: 92–8. https://doi.org/10.1051/lhb/20150060.

Durkheim, Emile. 2011. *De la division du travail social*. Editions Norph-Nop. Kindle edition.

Foltz, Bruce. 1995. *Inhabiting the Earth: Heidegger, Environmental Ethics, and the Metaphysics of Nature*. New York: Humanity Books.

Heidegger, Martin. 1980. *Gesaumtaufgabe 2, Band 39: Hölderlin's Hymnen "Germanien" und "Der Rhein."* Frankfurt: Vittorio Klosterman.

– 1993. "The End of Philosophy and the Task of Thinking." In *Basic Writings*, edited by David F. Krell, 431–49. Oxford: Routledge.

– 1995. *Being and Time*. Translated by John Macquarrie and Edward Robinson. Oxford: Blackwell.

– 1995. *The Fundamental Concepts of Metaphysics: World, Finitude, Solitude*. Translated by William McNeill and Nicholas Walker. Indianapolis: Indiana University Press.

– 1996. *Hölderlin's Hymn "The Ister."* Translated by William McNeill and Julia Davis. Indianapolis: Indiana University Press.

Himanka, Juha. 2005. "Husserl's Argumentation for the Pre-Copernican View of the Earth." *The Review of Metaphysics* 58 no. 3: 621–44.

Hobbes, Thomas. 1651. *Leviathan or the Matter, Form & Power of a Commonwealth Ecclesiastical and Civill*. London: Andrew Cooke. Kindle edition.

Hublin, Jean-Jacques. 2008. *Quand d'autres hommes peuplaient la terre*. Paris: Flammarion.

Hugo, Victor. 1862. Les Misérables. Édition Libre.

Husserl, Edmund. 1970. *Crisis of European Philosophy and Transcendental Phenomenology: An Introduction to Phenomenological Philosophy*. Evanston: Northwestern University Press.

– 1981. *Foundational Investigations of the Phenomenological Origin of the Spatiality of Nature, Shorter Works*. Edited by Peter McCormick and Frederick Elliston. Indiana: University of Notre Dame Press.

Illich, Ivan. 1986. *H_2O and the Waters of Forgetfulness*. London: Marion Boyars Publishers.

John of Salisbury. 1990. *Policraticus: Of the Frivolities of Courtiers and the Footprints of Philosophers*. Translated by Cary J. Nederman. Cambridge: Cambridge University Press.

Levinas, Emmanuel. 1961. *Totalité et infini: Essai sur l'extériorité.* The Hague: Martinus Nijhoff.

Mithen, Steven. 1996. *The Prehistory of the Mind: A Search for the Origins of Art, Religion and Science.* London: Thames and Hudson.

– 2006. *The Singing Neanderthals: The Origins of Music, Language, Mind and Body.* London: Wiedenfeld and Nicolson.

Picq, Pascal. 2009. *Au commencement était l'homme.* Paris: Odile Jacob.

Rousseau, Jean-Jacques. 2007. "Discours sur l'économie politique." In *Du contrat social,* edited by Robert Derathé. Paris: Gallimard.

Saint-Simon, Claude Henri. 1965. *De la physiologie sociale: Œuvres choisies (extraits 1803–1825),* edited by Georges Gurvitch. Paris: Presses Universitaires de France.

Schuiten, Luc. 2010. Vers une cité végétale. Wavre: Madraga.

Sloterdijk, Peter. 2017. *Not Saved: Essays after Heidegger.* Translated by Ian A. Moore and Chirstopher Turner. Cambridge: Polity.

Stringer, Christopher, and Robin McKie. 1996. *African Exodus: The Origins of Modern Humanity.* New York: Henry Holt.

Todd, Nancy Jack, and John Todd. 1993. *From Ecocities to Living Machines: Principles of Ecological Design.* Berkeley: North Atlantic Books.

Varela, Francisco. 1996. "Neurophenomenology: A Methodological Remedy for the Hard Problem." *Journal of Consciousness Studies* 3 no. 4: 330–50.

Ward, Frederick O. 1856. "Discours prononcé à la séance d'ouverture du Congrès International de Bienfaisance" (Opening speech at the Congrès International de Bienfaisance), Brussels, Belgium, 15 September.

Zola, Emile. 1878. *Le ventre de Paris.* Édition Libre.

12 Towards a Complexity Ethics: Understanding and Action on Behalf of Life-World Well-Being

ROBERT MUGERAUER

Introduction

A 2015 UN report finds that ten years from now, we will have 40 per cent less water than we need. This adds to what we unavoidably already know: worldwide, 1 to 2 billion people lacked access to safe drinking water in 1990 – a figure that remained unchanged through 2000 (Myers and Kent 2005, 132). Today, over 2 billion people are reported to use a drinking water source contaminated with feces, and reliable projections suggest that by 2025, half of the world's population will be living in water-stressed areas (WHO 2019). Half the world's hospital beds are occupied by people suffering from water-borne diseases (United Nations Environment Programme 2015).

Moreover, this situation is connected to startling, indeed increasing, disparities in wealth and access to resources, in which 10 percent of the world's population – 736 million people – survive on less than US$1.90/day (World Bank 2019). In the words of the World Wildlife Fund, "[i]t's the poor who depend most on natural resources for their livelihoods, and who suffer most from the impacts of climate change, deforestation, overfishing and other environmental problems. Environment and development issues can't be separated (WWF 2019). Many people live in the midst of polluted air, with contaminated water that may have to be collected from sources close to factory outfall pipes and on soil that may be contaminated by lead, mercury, and chlorinated compounds" (Girardet 1992, 96). These and related water problems will only become worse when rising water levels due to climate change result in significant displacement of populations living on coastlines and on large rivers. (In 2017, 2.4 billion people, over 30 per cent of the world's population, lived within 100 kilometres of a coastline [United Nations 2017].)

Clearly, we are at the point where ethical concerns blend into politics, especially in public policies and practices that bear on the good of all (including the sustainability of the environment). Such water ethics require sensitivity to their contexts – the ethos of widely varied public–private social-economic structures around the world, including but not limited to government agencies, businesses and industries, and the public – as well as a focal concern with distributive justice.[1] As part of our general ethical responsibilities as people, as well as our professional obligations in practice and as educators, researchers, policy makers, private sector planners, and environmental professionals, in dealing with water ethics we need to consider who ultimately benefits and who suffers, as well as how to involve inhabitants in participating more in decision making, if not actually making decision making more democratic.[2]

Attending to how we currently attempt to understand and act appropriately in our world, it becomes immediately apparent that the sciences, especially with the advances in the bio/life-sciences, have provided fresh and powerful ways to think and act. In contrast, the humanities-based axiology that had traditionally been important in dealing with our realms of meaning and value has either faded in perceived importance or, where still significant, has too often not had practical outcomes. This is certainly true of ethics, both within our personal lives and in conjunction with politics and public policy, which bear on our shared lives. This is not to argue that the traditional approaches of natural law, deontology, or utilitarianism do not continue to be important but that attempts to improve our theories and practices remains a permanent responsibility for everyone. Hence the move to supplement or replace the principle-based ethics of the traditional views with broader, more holistic approaches: for example, as originally motivated by feminism, with an ethics of care (which in turn enriches our thoughtfulness, such as by having to re-examine the related virtue ethics). In attempting to describe, interpret, and analyse environmental-social issues raised by cases of waterway development and remediation, I have found that a further step would be well warranted. Developing a version of *phronesis* congruent with complexity theory, self-organization, and enactivism would enable us to develop an ethics and politics able to better deal with complex, nested environmental-social problems by benefitting from the sophisticated treatment of the multi-levelled dimensions of life that these sciences are fruitfully unfolding. Because what I have written is the outcome of reflecting on what emerges from an empirical investigation of concrete cases, this chapter will not proceed to "demonstrate" the correctness of the posited view. Rather, it will explore a case in adequate depth – Geertz's thick description – so that the above points can be seen arising from the situation, such that though they do come from a particular situation, they generally resonate with many related cases elsewhere. Insofar as this is successful, the chapter will then make a fuller account of the characteristics of an ethics of life-world complexity.

The Case: Lower Duwamish Waterway

The Basic Situation

As a phenomenon in the world, and not just a trope organizing our reflections, rivers frequently present multiple (often contradictory) ethical environmental-social issues at the same time, ranging in both spatial and temporal scale. One such case involves the Duwamish River, which runs down from Washington's Cascade Mountains to Puget Sound and Seattle, located at the point of outflow. With the Duwamish River (or the Lower Duwamish Waterway, to give the place its proper designation) we are given a tangle of historically unfolding claims for use that involve significant harms to some groups and the environment common to all, and thus issues of justice and appropriate future actions for rehabilitation. The initial inhabitants of this section of the Pacific Northwest were displaced by the first European settlers who farmed the rich bottom land along the river's course. As Seattle grew through industrialization, the river and atmosphere were increasingly polluted for over a hundred years, especially through the military production of the Second World War, with the heavy manufacturing of Boeing and other companies. Given the careless discharge of toxic materials into the river, with no end in sight, the area was declared a Superfund National Priorities site in 2001, with a high priority and mandate for clean-up. A thread of continuity, through the disastrous situation where at best a modest, incremental recovery to a "tolerable" level of pollution is possible, but never a return to genuine high-quality water, appears with tribal fishing rights and practices. Though suppressed significantly over the last one hundred years, tribal members have recently had their rights recognized, both through the courts and a more enlightened citizenry; concurrently, they are increasingly active participants in river activity and dealing with the current problems. The waterway, Seattle's only river, is the site of annual salmon migration, which is critical given that the fish have been a major source of food and income for the tribes from the beginning and are still an important dietary and symbolic dimension of their lives. The water is so polluted that eating any of the river's "immobile" inhabitants – bottom-feeding fish, clams – is counter-indicated.

The reaffirmed fishing rights and findings by health agencies do, however, allow for salmon to be legally and safely caught during their river runs – the time of pass-through from the ocean and then the sound upstream is short enough that the salmon, which do not dwell in the waterway itself, are safe to eat. Importantly, the current dynamic results from both the tribal fishing rights and the fact that it is safe for salmon, which has become the regional totem (mainly through the enthusiasm of the active ecological community). The legal apparatus of environmental protection for the salmon, coinciding with the exercise of fishing rights, now generates and supports the move to mitigate the environmental damage.

Though this limited catch is able to continue, the heavily polluted waters remain a significant problem because many people, mainly non-Native Americans, continue to fish from the banks, either because they are unaware of or unable to read the many warning signs, or because they are poor and these fish and shellfish are a crucial part of their diet. At the least, this raises the question of the extent to which the public sphere has an obligation to provide an alternative food source. There have been proposals to make safe-to-eat fish available, but it is not clear how these could be implemented. Programs to swap a safe for a non-safe river catch will not work, because that would be a push to continue fishing. Nor will providing vouchers for safe fish to be redeemed at local grocery stores work, because there is no clearly defined, stable set of people who now "improperly" fish: identifying and targeting those in actual need as documented by their recent activity would be impossible. How, then, even within a clearly circumscribed sphere – the inedible river dwellers and a clear responsibility to prevent further harms – can something practical be done? What good is settling ethical responsibilities, calculating the degrees of risks of harm, if there is no actual outcome? The problem will remain because there is general agreement that, no matter how much remediation cleans up the river, the water will never be safe to swim in or drink; the fish will never be safe to eat.

Recognizing the limits of dealing with the pollution of the water itself, which must remain underway, attention shifts beyond the river proper to the banks and further inland to the communities of Georgetown on one side and South Park on the other. The shoreline, which could be an amenity, is in fact a threat more regularly encountered than the water itself. The soil is so heavily contaminated that it is deemed unhealthy for children to play there. The poisonous effect, then, spreads upward, through an especially dangerous zone, to the rest of the community.

The neighbourhoods of Georgetown on one side and South Park on the other contain a mixture of housing and industrial-shipping activity. Setting aside the responsibility and liability of the latter for generating the high levels of contamination, even the legal and political recognition that has led to the current efforts of reclamation, has effectively come to a stop – hopefully, only a halt in what might be done next. That is, ethical issues arching from the past to today yield even more complex problems for the future. Since there is no prospect of remediating the entire area (especially the water) to a really good state, the harms already done appear to open continuing environmental injustices. Importantly, the burdensome damage resulting from the environmentally irresponsible activities manifest in the polluted water and shoreline eventually affects the overall character of the neighbourhoods involved because of a complex of feedback loops in the physical and social systems. The *South Park Green Space Vision Plan* (2014) summarizes the situation:

> However, South Park also suffers from environmental inequities such as lower than average life expectancies and less public green space than other parts of the city. The Duwamish Valley Cumulative Health Impacts Analysis shows that people living in the Duwamish Valley are exposed to more pollution and live, on average, lives that are eight years shorter than residents in other parts of the city. And, while it's a well-known fact that access to parks, trails and healthy recreational opportunities correlate to improved health and happy citizens, an average of only 40 square feet of accessible open space is available to residents of South Park, versus the average of 387 square feet per resident within Seattle City limits and up to 1100 square feet per residents of some wealthier neighborhoods. (2-2)

For example, South Park residents' life expectancy is eight years shorter than the county average and thirteen years shorter than in Laurelhurst, a prosperous neighbourhood on the protected shore of the inland Lake Washington; heart disease rates in South Park are 47 per cent higher than the county average (HIA 2013, 15; Gould and Cummings 2013). Approaching an adequate response would involve clearly understanding the interactions among the many dimensions, where negative features dramatically multiply in a downward spiral, though not in a simple linear causal sequence. The bad biochemical environment, through a complex of positive and negative feedback loops, is the scene of a high crime rate, high unemployment, low property values, and marginal infrastructure and services.

Starting from the pragmatic, the ethical issues of resource allocation and resident autonomy are ever more pressing because of the inadequacy of available resources. Decisions are needed concerning both the short and long term as to how to maximize the few likely benefits received from legally mandated and politically responsive sources. Each apparent, and at least partial, "success" carries within it choices among possible prospects, the weighing of benefits of what kind for whom of qualitatively different projects. Utilitarian ethical theory has little to offer here, as becomes even clearer in light of the highly diverse values and views of participants who would legitimately contribute to the discussion.

There are difficulties to resolve in regard to fair representation. A small number of strong voices are able to convey the concerns and viewpoints of groups that are at least semi-organized (for example, neighbourhood associations) and especially for organized non-profit organizations or government agencies. This becomes problematic insofar as participants include a significant number of people who have the time to regularly present their positions, while many community members are not able to do so because of work or family obligations. This raises the structural question – a variation of "fair opportunity" – of how to ameliorate the uneven access to civic processes by those with higher income levels or by those with higher formal status who come as recognized representatives of non-governmental organizations (NGOs). Even if they are well intended and have valid ideas, insofar as they claim to speak for many, if not all (which is not the case), residents find themselves pushed aside and begin to resent these "loud voices" who have power disproportionate to their numbers. This uneven access by contributors operates not only in discussions among the residents, but in an even more dangerous manner in the crucial interactions between residents and governmental or corporate sectors. The latter transactions become even more aggravating to ordinary inhabitants when, as regularly happens, meetings at which higher-level decisions are discussed, if not made final, are scheduled in downtown or suburban offices during the day.

Thus there is a cascade of fractures that run in many directions. How could the right to autonomy be exercised? This would not be just a question of having their informed views heard and then having them either prevail or give way (by legitimate decision of some kind, vote, etc.), but of unequal ability to speak and be heard in the first place, to have a process that continues along with the appearance of propriety but where unauthorized "proxies" carry their positions forward. (Of course, this is most likely well meaning, so clearly a matter of paternalism; if not, then there exists an even deeper problem of usurpation, of suppressing residents' autonomy.)

As to the actual, though thus far meagre, projects that could benefit South Park and create a better quality of life, to try to begin to turn the downward spiral around, multiple meetings have been held to allow residents to say what they would like to see changed, to interact with and sometimes be guided by many groups of environmentally related professions or other sorts of "experts" (open space and park planners, designers, bicycle advocacy groups, official transportation specialists, numerous city and county government staff, and plenty of consultants). One of the major efforts resulted in the *South Park Green Space Vision Plan* (2014). The result was a set of recommendations for partnership opportunities, funding sources, and priority sites to improve over the next five years. Top priority sites identified by the community include

1 South Park Community Center
2 Duwamish Waterway Park
3 Shoreline Street Ends
4 14th Avenue Corridor between S. Henderson St. and Dallas Ave. S.
5 South Park Plaza
6 Concord Elementary
7 Walking and biking connections between all of the above. (1-1)

Admittedly, the elements were part of a specifically Open Space/ Green Space plan; nonetheless, the atomization of thinking here is all too apparent. Though the ostensible goal of the project was to "improve the health of the environment and community" (1-1), the horizon was limited to the benefits of open public spaces, with no attention to how its residents interacted with the numerous other dimensions of South Park's diminished quality of life. In this sense, the professionals were acting ethically enough by normal standards – seeking to decrease harms and ameliorate some dimensions of environmental justice; but as often happens, these separated items were here considered an adequate "solution" for a broad range of unmet needs, with only fundraising and implementation remaining (not that it would likely have happened anytime soon). Though more could be said about the autonomy of the stakeholders (exploring questions of paternalism, informed consent, and so on, in the decision process), the above should make clear that despite modest public participation (the *Plan* lists sixteen active participants *ex officio*, eight consultants, and six "community members"), we can see that issues of priorities in the allocation of resources were decided on the basis of the power of NGOs and local government agencies. In essence, the people consistently and successfully concerned with quality of water attended to water; those concerned with parks attended to parks.

Towards a Fuller View and More Adequate Response

Because it does not focus on the overall life-world with its complex interactive dimensions, the process just described would be, perhaps, agreeable enough, but would certainly not provide an adequate approach to the real problems. One way to put the point is to question what the professional helpers are responsible for. To see that their area of expertise is covered? Or to see to the health of the whole? Thus, while not unethical, the professionals are working within a much too limited conception of ethics. This can be seen in regard to two dimensions. First, the actual empirical phenomena are largely ignored: Would a urologist be acting properly if she entered the operating room, removed a kidney stone, then was not seen again? The medical team recognizes that what is happening with the kidneys – for instance, in the case of hyperparathyroidism and a calcium imbalance – may also involve many other factors, such as decreasing bone mass. The kidneys are but one part of the body, interactive with the rest of our organism, so that the medical team is responsible for the entire homeostasis, and thence the well-being of the person. Second, an adequate ethics would be a matter of the *group's* responsibility for the whole. Otherwise, how could we even begin to puzzle out the basic issues: the tensions and trade-offs of benefits and harms that must be understood before we can weigh them and decide, or who is centrally affected among them, in the case where only the mother or the about-to-be-born child can live, or there is only enough vaccine for one small segment of those who need it (how desperately? with what lasting effects?). As these simple examples make clear, the whole is not limitable to a person, but to persons in relation to others they may neither know nor ever meet.

The view I am developing has a good deal in common with some versions of feminist ethics of care, which initiated the drive to a fuller view of who is involved in discussions of moral considerability and in what relationships (see Hamington 2004). Some similarities to an ethic of care – at least those who find that such is compatible with a justice-based ethics (see Card 1991 and Oken 1989)[3] – lie neither in focusing on the systems of rationalized principles, as with natural law, deontology, and utilitarianism, nor so much on moral judgment and justifications, but on conduct, "concrete elements of situations," and relationships beyond the autonomous individual as the "unit of authority" in making judgments. Such broader concerns with embodied care, beyond narrow rationality as conceived in the modern era, importantly include attention to habits, knowledge more broadly, and imagination (and thus interesting connections with phenomenology).

I will argue, however, that 1) what might be called motives of action, 2) the complex character of the interactive networks of relationships to be considered,[4] and 3) the details of the concrete phenomena require more than an ethics of care has provided thus far. In regard to the first aspect, rather than focusing on "feeling and sentiment" as deep motivations, or on "clusters of principles revolving around care ... and trust" (Manzo 2008), a more multidimensional appreciation of interactive factors is required, interpreted from a life-world trajectory to realize values ranging from the functional to the vital and cultural (and perhaps spiritual) as participants in a concrete, historical ethos (Scheler 1973). In regard to aspects 2 and 3 (about which I will say more shortly), the approach developed below is closer to virtue ethics in looking at dispositions, particularly when they are understood as complex, and take into consideration a broad range of factors and conditions.[5] The case above illustrates how essential this is if the intricacies of a community's ills and possible well-being are to be adequately understood and an extended, heterogeneous group gathered to find a way to conduct themselves – a position indirectly developed in Maturana and Varela's (1992) ideas of maintaining organizational stability by way of the self-adjusting reciprocal and re-iterating structural couplings between a person and their social-natural *Umwelt* – which can be interpreted as a bio-phenomenological way of articulating *phronesis*.[6]

An ethics of complexity, then, is concerned with conduct in existential situations and networks of relationships (including the assemblage of factors bearing on the multiple networks active among persons, such as informal groups, formal organizations, things, events – the hierarchical feedbacks emerging from and constraining biochemical, biological, physiological, pre/sub-personal, personal, cultural-political, and large-scale ecological systems) whose positive and negative feedback loops can lead towards or away from the particular constellation's social-ecological well-being.

But how can this more robust understanding come about? It would be as a combination of empirical analysis able to deal with both individual elements and their complex interactions and of an integrative, if not holistic, overview or framework. Both of these features are characteristics of the cluster of recent sciences dealing with open, non-linear processes, autopoietic systems, developmental systems theory (DST), dialectical biology, and neuro-phenomenological enaction – which often, for convenience's sake, thus setting aside the more precise differences, are referred to as "complexity theory" or self-organization.[7] In the simplest form, these sciences convincingly investigate distinct empirical phenomena to deal with the multi-causality (feedback loops) operative among them, in part by clarifying the hierarchy of emergent levels (a current, sophisticated update of views first developed by philosophical anthropology). Figure 12.1 shows a simplification, using accepted non-arbitrary categories arrayed in a set of hierarchical levels of order of processes or emergent phases.

Macro/Cosmic

World

Biosphere:

Atmosphere, hydrosphere, lithosphere

↓ ↑

Bio-cultural-regions

↓ ↑

Ecosystems of communities

↓ ↑

Communities (of populations of organisms)

↓ ↑

Umwelt/lifeworlds

↓ ↑   **Human lifeworld,** including built environments

↓ ↑   3. Political–ethical life

↓ ↑   2. Unique individuality of each person

↓ ↑   1. Specifically human mode of embodiment

↓ ↑   (b) Intentional movement and action

↓ ↑   (a) Sub-personal physiological–neurological–psychological

Organisms

↓ ↑

Organs–immune, nervous, and endocrine systems

↓ ↑

Membranes

↓ ↑

Bio-chemical/molecular (genetic)

↓ ↑

Micro/Sub-atomic

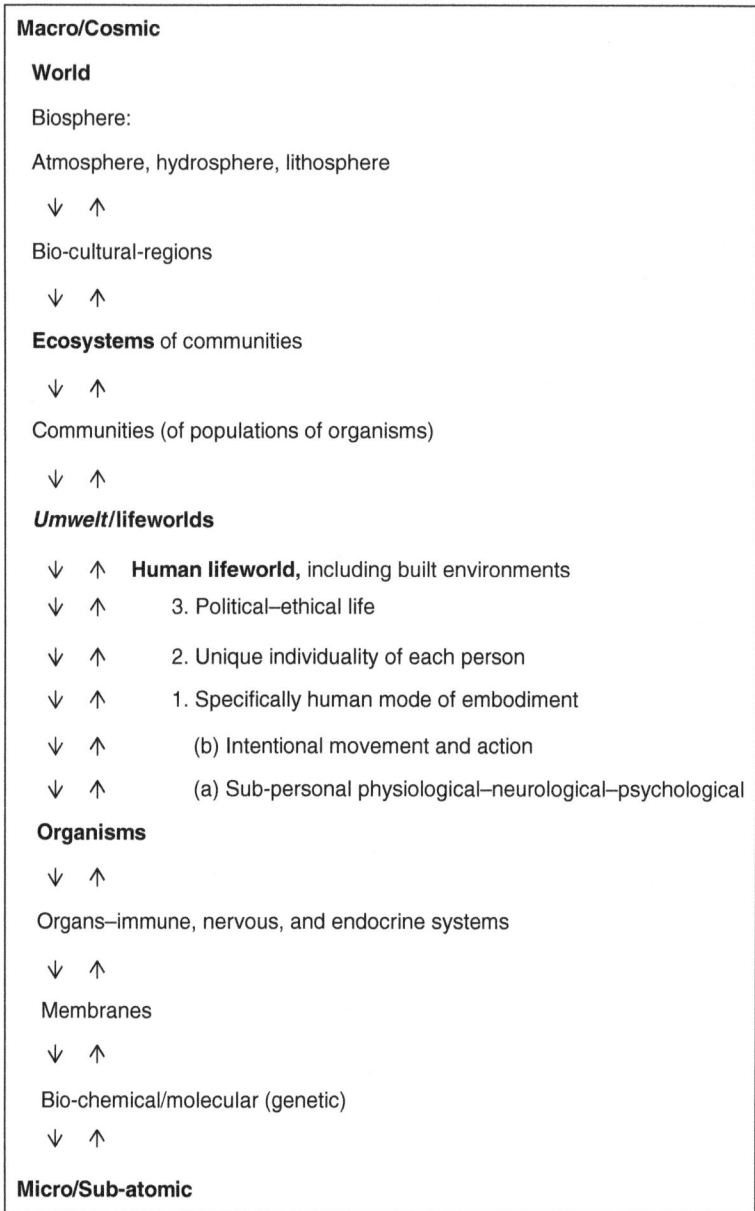

Figure 12.1. Anatomy of life and environmental interpretation (after Mugerauer 2010)

The powerful implication this has for ethics is that basic evaluations of beneficence, non-maleficence, autonomy, and justice cannot be made as simply as they usually are. While this is obvious in the case of medical ethics, where the action taken in relation to one biochemical dimension clearly reverberates throughout the organism and thence to the social realm, the same is just as true of any environmental ethics (though there is even more to water ethics than that). We cannot remain at the level of debating who is responsible for the pollution of water from a factory, much less, with a similar restricted concern, engage adequately the tangle of who is harmed by such pollution and how strongly without taking into account the web of relationships outlined here. PCBs, cPAHs, arsenic, and dioxins/furans in the sediment and water would affect a child in many ways, since changes in one element at a given level are not only operative at that level, but have emergent features. Beyond fundamental medical harms (cancer or damage to cardiovascular, neurological, liver, or immunization systems), these toxins may result in developmental problems (HIA 2013, 10), likely requiring extensive educational support. They may also involve a substantial drain on the emotional and financial resources of the family (and likely chronic stress, which negatively affects their mental health); prevent active participation in employment and thus non-contribution to generating more resources; and so on. This dense system of feedback loops includes all the dimensions shown in Figure 12.1.

Here we need to consider an extended example. The reader may object that even the next five paragraphs are too time consuming to spend time on, or are unnecessary, as many feel about long sections describing whales and gear in *Moby Dick*. On the contrary, our impatience with extended, more detailed description, especially of empirical detail (rather than the more obviously juicy matters of social and political intrigue), is a major symptom of the problem facing environmental ethics. What can we expect if we demand quick decisions concerning incredibly complicated phenomena, many of which operate on a scale of geological time rather than that of our contemporary attention span? An ethics of complexity would insist on patience with empirical detail.

To at least start to understand residents' reported concerns with safety and the crime level in this neighbourhood of poverty and poor environment requires making explicit the linked multi-causal processes operative across levels of the hierarchy of feedback loops, especially how the biochemical and physiological elements generate the personal and social. Consider the phenomenon of being attacked or even seriously threatened – say, in a narrow, dark alley. In the upward dynamic that occurs when we immediately experience a place as dangerous or scary, initially our sensory systems generate sub-personal biochemical and physiological responses as an anticipatory response to unpleasant fulfillments. Smells, for example, can move us from distress to anguish, from fear to terror (Colombetti 2014, 35–8, 40–4; Brennan 2004, 70). Dissonant sounds, such as can come from disturbed garbage cans or dropped metal echoing off the walls, will generated unease and perhaps even fear, which would combine with the visual impression of their source and import.

Reading the expressions of those around us – their facial, bodily, and vocal behaviour – will induce further feelings. If we encounter threatening figures, we may become fearful or angry. In what people are saying around us, not just the content but the aural rhythm and voice tones affect us: their anger may shake us up. Or, finding someone else already there, afraid and crying, maybe even hurt, could engender a rush of empathy and complicate our initial impulse as to what to do next. Our experience is intertwined with the emotions of others involved in the scene, since emotions are certainly transferred – offloaded – onto others. Those who receive the emotional impact have to cope with these emotions one way or another: by accepting them, by incorporating them, by resisting them. So, encountering the fearful situation means we have "taken on a disturbance," the emotional disturbance of the others, and we "have to adjust to the disequilibrium" (Brennan 2004, 29–30). How we do so is open, though likely entrained by our personal history, our disposition at the moment, and many other factors: our initial fear or anger can transform into an urge to flee or to turn aggressively on the source of the discomfort; response to a related helpless victim further confuses what we might feel and how we might act (Colombetti 2014, 26; Brennan 2004, 30).

As to the biochemical processes in stressful situations such as our scary alley, our autonomic nervous system activates the specific sub-class of stress-glucocorticoids. The sympathetic nervous system, the source of responses such as "vigilance, arousal, activation, and mobilization," will generate epinephrine (usually called adrenaline), which is in the same steroid class of hormones as estrogen and testosterone. At the same time, men's "testosterone is elevated in relation to the aggressive" situation, so that with an additional physical-chemical effect we feel "anger and tension" (which, via the feedback loops, further "elevate our testosterone and aggressive propensity"). In events where aggression leads to reciprocal hostility, there appears to operate a "coercive attraction into which participants are pulled that is not deliberately established, but that unfolds from moment to moment" (Brennan 2004, 71, 77–80; Colombetti 2014, 68).

Physiologically, the stress situation involves a "contraction in skeletal muscles; the hypothalamus sends chemicals to the pituitary gland which prompt, as just noted, the release of a range of hormones into the bloodstream (and, it is hypothesized, perhaps released into the atmosphere); breathing becomes faster and deeper; blood pressure rises; throat and nostril muscles expand their passages; perspiration increases" (Brennan 2004, 165). Of course, while much is going on here at the sub-personal unconscious level, we can consciously sense many of these changes: as a person with a past set of experiences and reactions and dispositions, as just noted, we interpret the bodily display we see and the voice changes we hear, as well as what we smell and touch. The effect of the environmental impact and the persistence of our resultant affects result from this combination of what we now experience from the environment and our history. That is, here we are engaged in a fully personal manner. It is somewhat easy to see how group violence is stimulated in the "circular reactions" that take place in each individual and across individuals into one or more groups. As a result of experiencing such situations, few alternatives are unproblematic: the stress might be internalized, though evidence is that withdrawal into oneself often leads to depression; externalizing the stress in the form of anger and aggression results in further continuation of hostility (Brennan 2004, 6, 43–4, 52, 59).

The emotions, as considered here, are fully social: they emerge in a world that we share with others in the neighbourhood; we display our fears and anxieties to each other; we react to others in ways that aim to reinforce, change, or suppress emotions in each other (*Mitsein*). Part of what makes residents' interactions in the neighbourhood important is that they can come to "recognize that another's feeling is the same as mine" insofar as they find a way to share intimacies – such as anger at being marginalized, or discouragement at being excluded from decision making that affects their lives; there is a basis for common action. In such solidarity, the neighbours are engaged in "participatory sense making," a realm where emotions are "socially extended" by the multiple reciprocal engagements generated and modulated, sometimes maintained, sometimes subtly or assertively ended (Colombetti 2014, 68, 172, 182). This is the world of complex social couplings by means of which the residents draw in many of the resources they need from each other and from interacting with surrounding natural and social environments. In sum, the biochemical and physiological processes generate – from which emerge – a suite of emotions, all of which may be reciprocally experienced so as to become assembled into shared social group experiences, within which occur multiple reciprocating behaviours – where none of these processes are linear or predictable.

To return to the particulars of the Duwamish Waterway, where would a complexity ethics leave us? A more comprehensive study – *Health Impact*

Assessment: Proposed Cleanup Plan for the Lower Duwamish Waterway Superfund Site (2013) – did conduct its research and articulated a plan from the complexity viewpoint (though it had not been so explicitly conceptualized or named). No doubt some of the broad, inclusive dimensions proceeded from the fact that the project was under the umbrella of the Environmental Protection Agency (EPA) (of course, a federal-level source does not in any way especially lead to broad and integrated coverage). The best explanation for the quality of the plan is that the multi-partnered study was mostly conducted by the Department of Environmental and Occupational Health Sciences of the School of Public Health at the University of Washington, particularly that at least two of the principal researchers do substantially practise in the mode I have described. Unfortunately, what had been commendably developed was set aside for, apparently, funding and political reasons. Nonetheless, some details are worth considering because they embody the substance, processes, and recommended actions that would be expected by a complexity ethics.

To maximize the ethical aspects of participation, equity, and informed consent, a very inclusive and extensive series of meetings involved all the stakeholders (the six major groups and agencies as well as numerous individuals representing the fishing community; the Suquamish and Duwamish Tribes as well as the Urban Indian Health Institute; the resident communities of South Park and Georgetown as well as homeless people and a member of the Governor's Panel on Health Disparities; and liaisons from fourteen private and public sector entities). Attention was given to the four groups that had suffered the greatest harms from the contamination: local residents, three Native American tribes, the highly diverse subsistence fishers (at least nine identifiable groups in addition to the tribes); and workers in local industries. Beyond preventing further harm from the clean-up, the goals also intended to maximize benefits and health equity – classical principles of ethical theory. The recommendations to the EPA, City of Seattle, King County, and Port of Seattle included careful measures for clean-up; truck and rail traffic; green remediation techniques; jobs for local firms and workers; attention to the highly contaminated sediments, resident fish and shellfish, and natural avian and mammal habitats;[8] unsafe stretches of shorelines; air, noise, and light pollution; stress and mental health problems; minimizing combined sewer overflow discharge; adding low-impact development stormwater systems and tree planting; improvements to streetscapes, access to the river, and safe open spaces.

Additionally, there was concern about the human levels of the hierarchy of co-constitutive elements: broader social and economic effects such as disruption of tribal traditions, loss of cultural opportunities, disempowerment, loss of social capital, loss of family wage jobs and dis-posable income, displacement or relocation. Appropriate to maintain-ing the life-world of current residents, employers, and regular others, the *Plan* stressed the importance of keeping an eye on the goals neces-sary for the community's resilience: increasing community engagement to build social cohesion, preserving affordability and producing afforda-ble housing, promoting and protecting home ownership (for example, from rising tax liability and foreclosure), and, not least, finding strate-gies of revitalization. There was agreement that this needed to be done, but without the usual disproportionate benefit to higher-income resi-dents and the increasing threat of land speculation and gentrification that feeds on low property values combined with desired improvements, eventually displacing residents and local businesses and industries.

That the HIA developed an appropriate plan is demonstrated not only by the content of the report and spirit in which it was conducted and recommendations made, but also by the subsequent confirmation, anal-ysis, and evaluation by Jonathan Childers's "Engaging Environmental Democracy: Tracing the Impact of the Duwamish HIA: Evaluation via Interpretative Phenomenological Analysis" (2014). In an even further step, two of the key researchers who authored the HIA study continued the project with a shifted emphasis to the social sphere. Though efforts to continue improving the water were important, as noted above, HIA found that "The models of future river sediment and fish and shellfish tissue concentrations predict that the Plan's health-protective goals will not be fully achieved" (2013, 30). Accordingly, their project "After the HIA" began the task of developing the core resident-industry system, since this substantially drives the positive feedback loops from which the community's recovery towards resilience could emerge. Though it was begun with grant funding, that did not continue, nor did the earlier political support.

Unquestionably, neither the full HIA nor (even more so) the "After the HIA" initiative operated as reductive science or with restricted pro-fessional focus (all too common in the spheres of public health and built environment). Beyond the knowledge acquired during the process and its analysis, the practical decisions and recommendations made and the actions taken were neither brute pragmatism nor mere cost-benefit calculations. We see this, for example, in the emphasis on implement-ing a mode of connection and circulation within the community, and on giving high priority to establishing an explicit reciprocity between a) helping industry to become "greener," thus enabling it to be successful while remaining in the neighbourhood to provide jobs, and b) having local employment to enable residents to continue to afford to live in their homes and provide a pool of residents eager to work. Here the researchers' understanding and interpretation of the matters before them, which facilitated working out the bio-cultural dynamic, was, though not so named by them, an instance of complexity theory; further, the manner in which they acted out of a value system and mindset – out of a multidimensional disposition to consider a broad range of factors to consistently move towards a right way of community being – manifested a substantive moral character. That is, both the mode of knowledge and the interest-conduct displayed an expertise that it is more than legitimate to call *phronesis* – sound practical judgment, largely gained and refined by way of experience, which is able to discern the weighed importance of one factor in relation to others when making decisions and choosing courses of action. Saying this obviously uses the vocabulary of virtue ethics: the virtues (*arête*), practically known by and enacted as *phronesis* for the sake of *eudemonia* – here the well-being of the shared natural-social life-world.

At the same time, I believe that the additional empirical and theoretical aspects of complexity, self-organization, and enaction warrant considering an ethics of complexity as having a distinctive character.[9] Still, the point here is not to argue for the superiority of such an approach over virtue ethics, but to profit from a concrete case of a waterscape to unpack and explore environmental ethics. The big ethical issue is our stance and response when making decisions and acting in regard to the tangle of harms, goods, autonomy of all affected, and justice (just to list the classic areas of ethical focus), or in more focal terms for this natural-social environmental case, how to adequately treat problems involving allocations of resources, proxy decision makers, fair opportunity and several dimensions of class and racial equity. Or, more deeply in the language of complexity theory and self-organization, the task is understanding and properly modulating the underlying dynamics that generate and maintain the system in a stable state from which the flourishing (or diminishing) of life in a particular *Umwelt* emerges. In ordinary terms, this means appropriately interpreting and acting on the tangle of multiple, reciprocally co-constituting elements and processes that affect the well-being of a life-world.

Implications

It becomes clear that ethics, of water or any named phenomenal realm, cannot be a matter, as it often has been, of focus on the individual, neither as the one whose rights are in question nor as the authority who is making the ethical judgments. Ethical responsibility appears here as a communal matter of many who are concretely engaged together in the world, across the arc of life – that is, in the full inter-determining bio-physical, organic, and specifically human physiological, intentional, uniquely individual, and ethical-political dimensions. Hence my characterizing this approach in terms of an ethics of complexity and shared life-world (*in-der-welt-mitsein*).

The broad communal nature of such an ethics is underscored if we enquire how it would come about that decisions and actions to deal with the riverscape would be made not by professionals focusing only on their own specialty, or what might be taken to be their "narrow" realm of responsibility, but as generous both in terms of the diversity of factors to be considered and the richness of the dynamic from which a natural-social *Umwelt* emerges. How can such a dynamic be maintained or restored so that a life-world might flourish? How would such an attitude be acquired? Surely only through what has long been called moral education, as a critical dimension simultaneously manifesting and being continually cultivated within a shared view about what sort of people we should be, of what sort of relation to the earth and waters we should have. If we say that understanding the actual empirical detail across the whole hierarchy of elements and processes amounts to a vertical dimension to the necessary ethics, while the communal (rather than overly individualized) mode of thought and conduct is a horizontal dimension across the community, then the challenge is to integrate the two. Which would be one of the principal reasons for writing this chapter and engaging in dialogue with its readers – the educational project of this collection. Hopefully, it forms a part of what is required to cultivate the disposition to act spontaneously so as to internalize all these directions in one's life and as an image of the world to be cared for – a disposition that would come, out of one's self, as a participant in a social ethos.

NOTES

1 Environmental justice (EJ), also known as environmental equity, deals with inequalities associated with the burdens borne by the poor, racial minorities, women, or inhabitants of developing countries. It seeks to redress unequal protection, decision-making practices that systematically disadvantage groups or locate undesirable or dangerous uses and facilities in proximity to their communities (Roseland 1998,154). Thus, environmental justice addresses inequality associated with hazardous sites (e.g., from toxic waste or particulate matter) located in areas inhabited by racial minorities, the poor, the elderly, or children (Agyeman 2005). The dimensions include

procedural equity, social equity, geographic equity, inter-generational equity, inter-species equity. Increasingly, governments at many levels are making EJ part of their missions.

Here, issues become comprehensive in terms of being affected: because decisions concerning sustainability are decisions involving entire mixed communities, they are fundamentally questions of distributive justice, especially because the principle of autonomy requires full, active participation – if not more strongly final decision making and informed consent – of those involved in the lifeworld in question. Disparities in environmental quality and decision-making processes are among the major challenges facing environmental ethics. There are at least two focal problems: to identify and include those who are unjustly treated, and to find ways to increase the self-determination and informed consent of those adversely affected. And, of course, disproportionate harms. Typically, issues have involved the impact on two major groups: first, the local inhabitants of life-worlds that are controlled by the political, economic, or other power relations such that an entire mixed community within which they live is damaged, as are their lives in this bio-cultural environment; and second, the marginalized within any given society, such as the poor or racial-ethnic-religious minorities. Lately, with transregional damage from environmental or economic causes, or even global impacts such as climate change, more complex still. This makes clear why our project of water ethics cannot concern only Canada or the United States: with marginalized groups living in desperate conditions in the midst of the development that brings opportunity and prosperity to others, by any normal measure there is clearly environmental and economic injustice.

Distributive justice (DJ) is also in full play in many of the issues considered here. There is an enormously uneven distribution of goods and harms, with very few receiving a disproportionate amount of the former and a very large number bearing most of the latter (Mugerauer 1996, 353–66; Engel and Engel 1990). DJ recognizes that there are both benefits and costs associated with the manner in which we distribute and use goods. The idea of DJ is that both the costs (or burdens) and the benefits associated without our providing and using resources need to be fairly or equitably or legitimately spread across different societies; and, within societies, equitably spread among diverse members or groups of a given generation and, equitably, across generations. The principle of fair opportunity would modify distributive justice.

2 In regard to professional responsibility, the American Society of Landscape Architects' *Code of Environmental Ethics* contains the following in regard to water ethics:

Ethical Standards 1.13. Water resources should be used efficiently and allocated equitably; all forms of water pollution should be eliminated to maximize the availability of safe drinking water; land use should conserve water and related ecosystems to sustain both human communities and natural ecosystems.

ES1.14. The natural and cultural elements of waterways and their corridors should be protected through the systems of national, state, and local designation of rivers and greenways to facilitate their integrity and use by present and future generations.

ES1.15. Wetlands are essential to the quality of life and the well-being of the earth's ecosystems; wetland resources should be protected, conserved, rehabilitated, and enhanced; and careful site-specific development and management efforts should allow for compatible land use, while preserving the ongoing functions of wetland resources.

3 Opposing the care-versus-justice dichotomy, Rosemarie Tong (1998) argues that the ambiguities of care can be largely resolved by bringing it closer to virtue ethics.

4 Treatment of relationships is better in both virtue ethics and ethics of complexity, using Humberto Maturana and Francisco Varela, who also have better answers for the question of the character of such a set of relationships as well as, uniquely, an empirically adequate response to item 3. (See Maturana and Varela 1992.)

5 How a group might acquire such a disposition certainly depends on the social context, the ethos of an already virtuous community. As argued since the ancients, this would be a matter of moral education. In our case, this often occurs, either almost from scratch or as improved, in the course of working as professional teams in close connection to residents, where mutual concern for "doing the right thing" overall prevails. Maturana's approach to ethics emphasizes how, as a member of a social community, an individual's dispositions to actions arise from patterns of the emotions underlying human relations, which in turn give rise to reflection on one's actions and thus the possibility of responsibility (2004, 77, 196–208).

6 More needs to be said about actual historical events and relationships to elaborate networks among people (individual, multiple, simultaneously operative informal, local groups, and in organizations) in multi-causal interactions with the materials, processes, and value systems of the life-world. Here, though, the focus, as with complexity and self-organization, is on the underlying dynamic that promotes positive social interactions and structures from which resilience or transformation, and overall well-being (of the social-ecological system), can emerge.

7 Major sources, respectively, are Prigogine (1980); autopoietic systems, Maturana and Varela (1992); developmental systems theory, Oyama, Griffiths, and Gray (2001); dialectical biology, Levins and Lewontin (2007); and neurophenomenological enaction, Varela, Evan, and Rosch (1991).

8 An interesting complementary study of Puget Sound, which would empirically fill in some of the biological-social processes in the hierarchy sketched

above, has focused on the life systems and *Umwelts* in the water that are structurally coupled to built environment elements, is Wilson (2008).

9   I am happy to find after rereading Varela (1999) for the umpteenth time that Varela himself would endorse this position (amazing how one can forget what one knows yet still find that the internalized account actually is correct). At the same time that he develops his own views on ethics, especially focusing on enaction, distributed neuronal self-organization, and tenets from Buddhism, he makes explicit his close connection with wisdom traditions. On ethical expertise, he explicates both Aristotle (by way of Alaisdair MacIntyre) and the Confucian Mencius, who holds that "An action is fully virtuous only if it flows from an activated disposition" in response to specific situations (not via rules or procedures) (23–36). In connecting "natural dispositions" to his view that a person "embodies ethics like any expert embodies his know how": in concrete situations, spontaneous, genuinely ethical actions occur where "truly expert people act from extended inclinations" – which, in turn, are only possible insofar as we are "full participants in a community" (23–36).

REFERENCES

Agyeman, Julian. 2005. *Sustainable Communities and the Challenge of Environmental Justice.* New York: NYU Press.

American Society of Landscape Architects' *Code of Environmental Ethics.* Accessed 26 February 2018. https://www.asla.org/ContentDetail.aspx?id=4308/.

Brennan, Teresa. 2004. *The Transmission of Affect.* Ithaca: Cornell University Press.

Card, Claudia, ed. 1991. *Feminist Ethics.* Lawrence, KS: University Press of Kansas.

Childers, Jonathan. 2014. "Engaging Environmental Democracy: Tracing the Impact of the Duwamish HIA: Evaluation via Interpretative Phenomenological Analysis." Master's thesis, University of Washington.

Colombetti, Giovanna. 2014. *The Feeling Body: Affective Science Meets the Enactive Mind.* Cambridge: MIT Press.

Engel, J. Ronald, and Joan Gibb Engel. 1990. *Ethics of Environment and Development: Global Challenge, International Response.* Tucson: University of Arizona Press.

Girardet, Herbert. 1992. *The GAIA Atlas of Cities: New Directions for Sustainable Urban Living.* New York: Anchor Books.

Gould, Linn, and B.J. Cummings. 2013. *Duwamish Valley Cumulative Health Impacts Analysis: Seattle, Washington.* Accessed 26 February 2018. http://justhealthaction.org/wp-content/uploads/2013/03/Duwamish-Valley-Cumulative-Health-Impacts-Analysis-Seattle-WA.pdf/.

Hamington, Maurice. 2004. *Embodied Care: Jane Addams, Maurice Merleau-Ponty, and Feminist Ethics*. Urbana: University of Illinois Press.

Health Impact Assessment: Proposed Cleanup Plan for the Lower Duwamish Waterway Super Fund Site Final Report. 2013. Seattle: Department of Environmental and Occupational Health Sciences, School of Public Health, University of Washington. Accessed 29 February 2018. https://deohs.washington.edu/health-impact-assessment-duwamish-cleanup-plan/.

Levins, Richard, and Richard Lewontin. 2007. *The Dialectical Biologist*. Cambridge: Harvard University Press.

Manzo, Lynne. 2008. "Ethic of Care." In *Environmental Dilemmas: Ethical Decision Making*. Edited by Robert Mugerauer and Lynne Manzo. Lanham, NJ: Lexington Books, 149–55.

Maturana, Humberto R. 2004. *From Being to Doing: The Origins of the Biology of Cognition*. Heidelberg: Carl-Auer Verlag.

Maturana, Humberto R., and Francisco J. Varela. 1992. *The Tree of Knowledge: The Biological Roots of Human Understanding*. Boston: Shambhala.

Mugerauer, Robert. 2008. "Theories of Sustainability: Environmental Ethics, Mixed-Communities, and Compassion." In *Environmental Dilemmas: Ethical Decision Making*, edited by Robert Mugerauer and Lynne Manzo. Lanham, NJ: Lexington Books, 353–66.

– 2010. "Anatomy of Life and Well-Being: A Framework for the Contributions of Phenomenology and Complexity Theory," *International Journal of Qualitative Studies of Health & Well-Being* 5 no. 2:5097. https://doi.org/10.3402/qhw.v5i2.5097.

Myers, Norman, and Jennifer Kent, eds. 2005. *The New Atlas of Planet Management*. Berkeley: University of California Press.

Oken, Susan. 1989. *Justice, Gender, and Family*. New York: Basic Books.

Oyama, Susan, Paul Griffiths, and Russell Gray, eds. 2001. *Cycles of Contingency: Developmental Systems and Evolution*. Cambridge: MIT Press.

Prigogine, Ilya. 1980. *From Being to Becoming: Time and Complexity in the Physical Sciences*. San Francisco: W.H. Freeman.

Roseland, Mark. 1998. *Toward Sustainable Communities: Resources for Citizens and Their Governments*. Gabriola, BC: New Society Publishers.

Scheler, Max. 1973. *Formalism in Ethics and Non-Formal Ethics of Values*. Evanston, IL: Northwestern University Press.

South Park Green Space Vision Plan. 2014. Seattle: Seattle Parks Foundation.

Tong, Rosemarie. 1998. "The Ethics of Care: A Feminist Virtue Ethics of Care for Healthcare Practitioners." *Journal of Medicine and Philosophy* 23 no. 2: 131–52. https://doi.org/10.1076/jmep.23.2.131.8921.

United Nations Environment Programme, 3/22/15.

United Nations. "The Ocean Conference," New York, 5–9 June 2017. Accessed 25 July 2019 at https://www.un.org/sustainabledevelopment/wp-content/uploads/2017/05/Ocean-fact-sheet-package.pdf.

Varela, Francisco. 1999. *Ethical Know How: Action, Wisdom, and Cognition.* Stanford: Stanford University Press.

Varela, Francisco, Evan Thompson, and Eleanor Rosch. 1991. *The Embodied Mind: Cognitive Science and Human Experience.* Cambridge: MIT Press.

Wilson, Meriwether. 2008. *Environmental Change and Built Environments in the Marine Nearshore: A Comparative Analysis of the Development of Three Large-scale Marinas in Seattle, Washington from 1950–2007.* Dissertation: University of Washington.

World Bank. "Poverty." Accessed 22 July 2019. https://www.worldbank.org/en/topic/poverty/overview.

World Health Organization, "Drinking Water." Accessed 22 July 2019. https://www.who.int/news-room/fact-sheets/detail/drinking-water.

World Wildlife Fund, "Fighting Poverty and Making Development Sustainable." Accessed 22 July 2019. https://www.wwf.org.uk/what-we-do/projects/fighting-poverty-and-making-development-sustainable.

PART FOUR

Closing Reflections

Conclusion – Looking Forward: From Poetics to Praxis

... Little we see in Nature that is ours;
We have given our hearts away, a sordid boon!
This Sea that bares her bosom to the moon,
The winds that will be howling at all hours ...
For this, for everything, we are out of tune ...

– William Wordsworth (c. 1802), "Sonnet"

That we today remain "out of tune" with the natural world is evident. Whether through climate change impacts or polluted air and waters, a sense of balance and sustainability with the earth that nurtures us, continues to be evasive.

To be sure, much has been accomplished since 1972, when the landmark United Nations Conference on the Human Environment was convened in Stockholm, Sweden, and an international declaration containing twenty-six principles was signed to protect the environment. At the very least, environmental issues are no longer considered to be simply the concern of small contingents of activists. Principles of sustainability pervade both the private and public sectors in a way that was inconceivable decades ago.

Moreover, a parallel series of events has seen increasing attention paid to issues about water. Most recently, in December 2016, the United Nations General Assembly adopted a resolution to launch the International Decade of Water for Sustainable Development 2018–28. Aiming to mobilize global action, the decade brings renewed attention and commitment to providing safe water to all.

As we seek to develop new policies and practices to sustainably advance water security, the conversations will likely be technical, research will be data-driven, and prescriptive remedies will seek measurable, quantifiable targets and outcomes. Each of these aims is commendable and constitutes a requirement for genuine change in preserving and caring for our water sources.

However, such calculative approaches – while necessary – are not sufficient. Whether advancing engineering solutions or developing theoretical guidelines in the newly developing field of water ethics, the foundation for any action is the inspiration that comes from our recalling how water is foundational to our very existence as living beings.

Water is more than an abstract object of enquiry, a utilitarian resource, an economic commodity, or a topic for disengaged policy recommendations. Water is life-giving, life-affirming, life-sustaining. We too easily take for granted that vitality, essentiality, fundamentality to being with water in our everyday policy making and planning.

The authors of this book, each in their own way, take this forgotten givenness of water and bring its wonder to light. When we plan cities, let us keep in mind that water has its own way, its own paths, its own movement. Mindlessly choosing to build within floodplains ignores these deep realities, for instance. Or when we argue that the costs of providing clean water to remote Indigenous communities are too high, let us not forget that the human right to water is more than simply an abstract ethical ideal: it requires that we respect the significance of embodiment and its ontological ties to the life-giving capacity of clean, safe water for all.

To help us approach water through such a new, originative paradigm, we begin and end this volume with poetry selections, following the lead of phenomenologist Martin Heidegger, who reminds us of the importance of thinking not simply pragmatically, but poetically. In his words, "we dwell altogether unpoetically," even though "poetry and building belong together, each calling for the other" (Heidegger 1971, 227). To think poetically requires a different sensibility – one that is careful, discerning, open, and deliberative. Such thinking invites us to "become hearers ... The deliberating ones and the slow ones are for the first time the careful ones. Because they think of that which is written of in the poem, they are directed with the singer's care towards the mystery of the reserving proximity" (Heidegger 1949, 266–7).

However, we should not conclude that the reference to slow, careful decision making is, in principle, unable to keep pace with our fast-moving world. The fact is that the kind of thinking that Heidegger advocates here may privilege, in certain circumstances, wisdom and lived experience over long, convoluted technical calculations of consequences. It may privilege informed intuition, mood, and affectivity that precede objective constructs or overly narrow and utilitarian cost-benefit measures. It invites meaningful consultation with stakeholders who are affected by policies and practices, so that final decisions are better informed and anticipate future effects. It recognizes that decisions about water are not bounded by limits such as court cases but may reflect complex ongoing community concerns. And so it supports adaptive decision making through community engagement, collaboration, and dialogue among diverse voices, each of which deserves to be heard.

In the end, such originative thinking ensures that we take planning and policy making to heart in a way that reflects lived experience in a tangible, embodied way. If water is a shared "resource" to be managed, such management can only be meaningful when we realize that water is not the product of our own manufacture. The ontological givenness of water invites us to engage with issues of water policy and planning humbly, carefully, caringly and, ultimately, more wisely.

Poet Dilys Leman closes this collection, describing how the narratives of our lives play out always within the hidden backdrop of water. In this case, the Don River, which weaves its way through Toronto, Canada's largest city, recalls for us how water intertwines with our daily existence, sometimes explicitly, but more often implicitly. From the joy of skinny dipping to the definition of cultural mores around women, to our buried wastewater to the lure of water itself, the collection reminds us of how water's presence and absence define the structure of our spatial and historical relations with the world. To that extent, water policies and planning scenarios do no less than reflect who we are together and collectively.

It is that recognition and acknowledgement that takes the conversation around water ethics and policy making from theory to lived practice, from calculation to meditation. A drop of water on our faces as the rain washes and feeds the landscape around us; the joy of immersing ourselves in oceans and lakes that recall our watery origins; the taste of water that quenches a seemingly endless thirst – these are only a few moments in our daily lives that remind us of the belonging of water to our very way of being. Water security is not simply a utilitarian offering, but rather is a defining ontological need.

Ensuring that water policies grow out of a deep recognition of that need, of that essential belonging, is what will help to ensure that they remain humane and responsive to the gravity of lived experience. Only if we move from logic to an embodied awareness of the wonder of water will our water policies and decision-making strategies hold the promise of genuinely engaging with our world in a caring, discerning way.

REFERENCES

Heidegger, Martin. 1971. "... Poetically Man Dwells ..." In *Poetry, Language, Thought*, translated by Albert Hofstadter. New York: Harper & Row.
– 1949. "Remembrance of the Poet." In *Existence and Being*, translated by Werner Brock. Chicago: Henry Regnery Company.

The Lure of Water: Four Poems

DILYS LEMAN

Skinny Dipping on the Don River, circa 1909

Hey, social worker lady! We're just kids
having fun in the dark. Don't shoot us

bare-bummed like this. Maybe take a
picture of our really skinny backs?

Sure beats Friday nights at home.
Dad's fist in a wall. His dirty *hey-hey-hey*

with Mum. Afterwards, her lighting
candles to *the bless-ed deaf and dumb.*

I'd rather watch how the moon bends
to kiss the river better, turns its purple

bruises into silver & gold, so fish can
splash & leap, show off their fancy pearls.

Lady, you really want to know what it's
like at night? Sewing machines drilling

in your sleep. Dreams of spelling words
you still get wrong   *pub- lick   m- oral-s*

Man at the bridge pulled my hand
lower down. He said *mum's the word.*

Lady, there's a bull's-eye on my spine,
a bullet waiting. I heard the jingle

just before the fortune teller lady
switched her bracelet to my wrist.

Lady, don't ...

I am moonlight on the Don River.
King maggot licking it clean.

Great blue heron   tip-toeing
from your eye.

Figure C.1. Skinny-dipping on the Don River circa 1909 (City of Toronto Archives, Fonds 1244, Item 1797)

On River Fishing: Advice to Women

After the fifth month, the womb
may become confused or restless

in the presence of quick-flowing
water. Avoid putting your hands in

your mouth after immersing (Weil's disease)
& wrapping your line on a high voltage

cable. No poking at rats, puppies &
ponies. Assume they are truly

drowned. If you must cast at dusk,
avoid hooking a bat (rabies). Waving

a flashlight frightens the fish.
Sensitive women do not need to know

about gutting & scaling or your
quick-fry method. It only sickens them.

Do not dress like a man (slouching hat &
waders) unless you really want to

be asked to share a beer. Do not question
a gentleman's technique – how to cast,

reel in, humanely kill. The river pike
bite, steal hair clips, wedding rings.

Refrain from looking into their eyes.
It only encourages them.

Figure C.2. Women fishing (City of Toronto Archives, Fonds 1244, Item 167A)

The Lure of Water

Bottom of well   her dropping
down   deeper   What is

resignation   but the tongue
scraped of its own vocabulary

She composes animal song
in the dark   Sound vibrations

loosen granite   the circulatory
system in drowned trees

Gold-tinged newts come
to rest on her clavicle

& in the saltlick crease
between her breasts

Mud woman turning
                    turning

          shivering chrysalis

                    her wings beating
                        in your mouth

Baptisms for the Twenty-First Century

After three days, the rain stops.
But the water worriers still twitch

like divining rods. They run out into
the streets, fall to their knees on sewer

grates, ears to the ground to hear a river
sloshing below. Livid. Stirring up a curse:

Street drains clogged   Sewer mains backing up
                     Council chambers full of shit

The water worriers & City councillors arrange
to meet up. Planners & civil engineers sign on.

They renounce hot pools, golf memberships,
indoor rinks & lengthy showers. Everyone strips.

They slip into eco-baptismal swimwear, hold hands
& form a sidewalk chorus line, humming *Moon River,*

while a senior water worrier recites the 2000+ toxins.
Some insist on being the very first to lower themselves

into the sewer (a former creek) that runs below the city.
It will flush them, one by one, into Lake Ontario.

Portuguese water dogs are already stationed
at the shoreline, awaiting each arrival.

It may take a while. Never whining,
they will paddle around in circles.

Contributors

David Abram – cultural ecologist and geo-philosopher – is the author of *Becoming Animal: An Earthly Cosmology* and *The Spell of the Sensuous: Perception and Language in a More-than-Human World*. Described as "revolutionary" by the Los Angeles Times, as "daring" and "truly original" by *Science*, David's work has helped catalyze the emergence of several new disciplines, including the burgeoning field of ecopsychology. A recipient of the international Lannan Literary Award for Nonfiction, he recently held the Arne Naess Chair in Global Justice and Ecology at the University of Oslo. David is creative director of the Alliance for Wild Ethics (AWE), an organization dedicated to cultural metamorphosis through a rejuvenation of earth-based oral culture – the culture of face-to-face and face-to-place storytelling. He lives with his family in the foothills of the southern Rockies.

Bryan E. Bannon is associate professor of philosophy and director of the Environmental Studies and Sustainability program at Merrimack College. His research focuses on how conceptualizations of nature shape our normative responses to the environment. He is the author of *From Mastery to Mystery: A Phenomenological Foundation for an Environmental Ethic* (Ohio University Press, 2014) and is at work on a manuscript tentatively entitled *Being a Friend to Nature: Building a More Hopeful Relationship with the Earth*.

Henry Dicks is an environmental philosopher at Université Jean Moulin Lyon 3. His research interests include water policy, sustainable urbanism, and the philosophy of biomimicry.

Janet Donohoe is dean of the Honors College and professor of philosophy at the University of West Georgia. She is the author of several articles on phenomenology and place, as well as a book titled *Remembering Places* (Lexington Books, 2014), and has recently edited a volume, *Place and Phenomenology* (Rowman & Littlefield International, 2017).

Jeff Gessas is a PhD student at the University of North Texas where his primary research interests include environmental justice, critical theory, Indigenous philosophy, ecofeminism, critical pedagogy, and ancient philosophy. His current research explores structures and theories of power and the subaltern – in particular the work of Gramsci and Foucault.

Dr Trish Glazebrook is professor of philosophy at Washington State University. She writes on Heidegger, ecofeminism, ancient philosophy, philosophy of technology, environmental philosophy, and climate change. Her current research addresses climate impacts and adaptations by women subsistence farmers in Ghana. She also researches the military use of drones.

Stephan Harding is coordinator of the MSc program in Holistic Science at Schumacher College in the UK, and has been the College's resident ecologist and tutor since its inception in 1991. He holds a doctorate in behavioural ecology from the University of Oxford and is author of *Animate Earth: Science, Intuition and Gaia.*

Sarah J. King is the author of *Fishing in Contested Waters: Place and Community in Burnt Church/Esgenoôpetitj* (University of Toronto Press, 2014) and is an associate professor of liberal studies at Grand Valley State University, where she also teaches in the Environmental Studies and Religious Studies programs.

Irene J. Klaver is professor in philosophy at the University of North Texas and director of the Philosophy of Water Project. She works at the interface of social-political and cultural dimensions of water, with a special interest in environmental imagination and urban rivers. She is finishing a book about the Trinity River in North Texas, and working on a monograph called *Meandering, River Spheres and New Urbanism.* Dr Klaver was Water and Culture Advisor for UNESCO from 2008 to 2013, and co-director of the International Association for Environmental Philosophy from 2010 to 2014.

Dilys Leman is a freelance writer, editor, yoga teacher, and poet. Along with other Toronto poets, she teams up with Lost Rivers Toronto to offer river poetry walks along the Don and Humber rivers. McGill-Queen's

University Press published her full-length poetry collection, *The Winter Count* (2014).

Kirby Manià is an academic and poet who recently relocated from Johannesburg to Vancouver. She previously taught English at the University of the Witwatersrand (Wits). More recently, she taught an introductory environmental humanities breadth course at the Faculty of Environment at Simon Fraser University. She is currently teaching at the University of British Columbia. She holds an MA in Modern Literature and Culture from the University of York (UK) and a PhD in English from Wits. Her creative work has appeared in numerous print literary journals and online magazines, while her academic research has largely concentrated on city literature, gold rush imaginaries, and postcolonial ecocriticism.

Martin Lee Mueller is coordinating the creation of a new master's degree in environmental pedagogy at the Rudolf Steiner University College, Oslo, in collaboration with Schumacher College, UK. His book *Being Salmon, Being Human* was published in 2017 and has inspired a stage performance by the same name.

Bob Mugerauer is professor and dean emeritus in the Departments of Architecture, Urban Design, and Planning, and adjunct professor in Landscape Architecture and Anthropology at the University of Washington. His books include *Dwelling, Place, Environment* (co-edited with David Seamon, 1985), *Heidegger's Language and Thinking* (1988), *Interpretations on Behalf of Place* (1994), *Environmental Interpretations* (1996), *Heidegger and Homecoming* (2011), and *Responding to Loss* (2015). His current research applies continental thought and dynamic complexity to issues of health and well-being: "Towards a Theory of Integrated Urban Ecology" (2010), "The City: A Legacy of Organism-Environment Interaction at Every Scale" (2011), "Anatomy of Life & Well-Being: A Framework for the Contributions of Phenomenology and Complexity Theory" (2011), and "The Double-Gift: Place and Identity" (2017).

Stephen Smith is a professor in the Faculty of Education at Simon Fraser University. His academic work, informed by phenomenological theories and methodologies, pertains to matters of curricular and instructional practices in health education, and physical activity promotion. Illustrative publications are the 1997 SUNY book *Risk and Our Pedagogical Relation to Children: On the Playground and Beyond*, and various journal articles in outlets such as *Phenomenology and Pedagogy*, *Phenomenology and Practice*, and *Journal of Dance and Somatic Practices*. Recent work addresses relational dynamics with horses and other companion species. This scholarship is grounded in movement practices that include flow arts, circus arts, and the disciplines of horse riding.

Ingrid Leman Stefanovic is a professor and former dean of the Faculty of Environment at Simon Fraser University, Vancouver, Canada. She is also professor emerita, Department of Philosophy, University of Toronto, where she spent the majority of her career teaching and conducting research on how values and perceptions affect public policy, planning, and environmental decision making. Dr Stefanovic has served as executive co-director of the International Association for Environmental Philosophy and senior scholar at the Center for Humans and Nature, Chicago and New York. Recent books include *Safeguarding Our Common Future: Rethinking Sustainable Development* and the co-edited volume *The Natural City: Re-Envisioning the Built Environment*.

Index